Object-Based Image Analysis and Treaty Verification

AF762954

Object-Based Image Analysis and Treaty Verification

New Approaches in Remote Sensing – Applied to Nuclear Facilities in Iran

Sven Nussbaum
Systems Analysis and Technical Evaluation, Forschungszentrum Jülich, Germany

and

Gunter Menz
Department of Geography, University of Bonn, Germany

Springer

Authors

Sven Nussbaum
Systems Analysis and Technical Evaluation
Forschungszentrum Jülich
Germany

Gunter Menz
Department of Geography
University of Bonn
Germany

ISBN: 978-90-481-7777-6 e-ISBN: 978-1-4020-6961-1

© 2010 Springer Science+Business Media B.V.
No part of this work may be reproduced, stored in a retrieval system, or transmitted in any form or by any means, electronic, mechanical, photocopying, microfilming, recording or otherwise, without written permission from the Publisher, with the exception of any material supplied specifically for the purpose of being entered and executed on a computer system, for exclusive use by the purchaser of the work.

This report was prepared as an account of work sponsored by the Government of the Federal Republic of Germany, Federal Ministry of Economics and Technology, under contract no. 02 W 6243. The author(s) is (are) responsible for the contents of this report.

The facts and opinions expressed in this Work are those of the authors and not necessarily those of the publisher.

Printed on acid-free paper.

9 8 7 6 5 4 3 2 1

springer.com

Preface

With the launch of very high-resolution satellite sensors at the beginning of this decade (e.g. IKONOS, OrbView, QUICKBIRD) many important new remote sensing applications have been developed. Major progress has been demonstrated in the accurate detection, identification and analysis of small-scale, site-specific features, including critical commercial and industrial facilities such as nuclear power plants. According to IAEA Director-General and 2005 Nobel Peace Prize winner, Dr. Mohamed ElBaradei, the provision of the necessary information and guarantees for monitoring the Iranian nuclear programme is an urgent priority for the future. Today, satellite imagery analysis is an integral part of the International Atomic Energy Agency's (IAEA) safeguards and verifications efforts and high-resolution commercial satellite imagery is an important component of this process.

Several international studies have proven the usefulness and documented the applicability of remotely sensed data for site verification purposes. Recent important developments in satellite sensor capabilities and innovations in analytic techniques make it clear that this "hardware-software-complex" is changing and improving rapidly. This book documents the latest developments in satellite data analysis and comprehensively describes the process of extracting information from space-borne sensor data for site verification purposes.

New image analysis methods are required if analysts and decision-makers are to derive the maximum information contained within the newly available very high-resolution satellite data. Traditional pixel-oriented algorithms and analytic techniques cannot take full advantage of the increased spatial coherence of very high-resolution imagery. In particular, object-based procedures are much more appropriate for the detection and the classification of man-made structures. Employing object features such as scale, compactness, orientation and texture, in addition to spectral characteristics, significantly extends the applicability of remote sensing satellite image analysis.

This book describes recent progress in object-based image analysis and also presents new results in applying this type of analysis to site verification for nuclear non-proliferation. A comprehensive workflow – including innovative algorithms for object-based very high-resolution image pre-processing, feature extraction, change detection, classification and interpretation – is developed, applied and evaluated. The entire object-based analytic flow is demonstrated, using several case studies of

Iranian nuclear facilities. This book is aimed at professionals and experts in the field of remote sensing and image processing, as well as political leaders and decision makers.

The authors clearly demonstrate that the new object-based image analysis approach has great potential for site evaluation in general, as well as specific applicability to the process of nuclear verification. The technologies of object-based remote sensing and remotely based nuclear verification are undergoing continual development and improvement. This book documents that while both technologies are already utilized as powerful tools for solving today's nuclear verification problems, they also have great long-range potential for the future.

We would like to express our sincere thanks to all those who contributed towards the success of this book. Special thanks go to the Research Center Jülich, the University of Bonn, the Freiberg University of Mining and Technology, the GMOSS (Global Monitoring for Security and Stability) Network of Excellence and Prof. Dr. D. Klaus. We would also like to thank the GMOSS Network for financial assistance in the Production of this book.

Jülich, Germany
Bonn, Germany

Sven Nussbaum
Gunter Menz

Contents

List of Figures

List of Tables

Chapter 1
Introduction

To smash the little atom, mankind was intent. Now any day the atom may return the compliment[1]

In most countries worldwide, the nuclear fuel cycle continues to be of great significance for sustainable development. Nuclear facilities provide electricity and heat, are used for research and medical applications as well as many other purposes. However, socially acceptable applications are only possible if nuclear energy is used exclusively for peaceful purposes and if its technical safety is guaranteed. Against this background, a policy of non-proliferation of nuclear weapons has been developed since the early nineteen fifties. Through a wide range of national and international policies, a network of rights and duties on the part of states and international organizations has emerged which has become known as the non-proliferation (NP) regime. The centrepiece of the non-proliferation policy for nuclear weapons is the Treaty on the Non-Proliferation of Nuclear Weapons (Non-Proliferation Treaty, NPT).[2]

The *International Atomic Energy Agency* (IAEA) with its headquarters in Vienna is responsible for monitoring trade in nuclear fissile material and the fissile material inventory of all state parties. The objective of the IAEA's work is to ensure a reliable, safe and peaceful use of nuclear technologies.

At the General Conference of the IAEA in September 2005, Dr. ElBaradei, Director-General of the IAEA and winner of the 2005 Nobel Peace Prize, put forward the following aspects as urgent priority challenges for the future:

- a general propagation of the additional protocol;
- expansion of the so called integrated safeguards;
- normalization of safeguards in Iraq;
- reintegration of the Democratic People's Republic of Korea (North Korea) into the NP regime;
- **provision of the necessary guarantees and information for monitoring the Iranian nuclear programme;**
- and further investigations of the illegal procurement network.

[1] Max Born c.1882 - c.1970.

[2] The complete text of the treaty can be found in INFCIRC/140 [IAEA/Hg., 1970].

S. Nussbaum, G. Menz, *Object-Based Image Analysis and Treaty Verification*,
© Springer Science+Business Media B.V. 2008

In summary, according to Dr. ElBaradei the IAEA verification system has to date displayed great flexibility and ingenuity in dealing with many of these challenges. Intensive verifications efforts have been speedily initiated in several countries and there have been extensive investigations concerning the illegal procurement network. The verification and safeguards systems have been enhanced by the increased application of satellites, environmental samples and a wide range of new technologies – as well as by the development of improved techniques for information analysis, the introduction of integrated safeguards and the transition to a more qualitatively oriented information system.[3]

The brief statement quoted above already shows the necessity and relevance of the topic of this book. The results of this work, with respect to both method and content, can make a contribution to approaching and tackling one of the major challenges of our time in the field of nuclear verification since information and reliable evidence on the Iranian nuclear programme are urgently required. Acquiring this information on the basis of satellite data is thus in accordance with the IAEA's approach since they regard the increased use of satellites as an enhancement of the verification and safeguards systems. Jacques G. Baute, Director of the IAEA's Iraq Nuclear Verification Office, can be quoted on the use of high-resolution Research & Development (R&D) data as saying:

> In the absence of inspections, high-resolution commercial satellite imagery provided the only real opportunity to remain in contact with the reality in the field. (Baute 2006)

As a complement to the methods of visual and pixel-based image interpretation, this work is concerned with the potential of an object-based image analysis for assisting verification efforts.

In the following chapters introducing the topic, the general field of safeguards and verification will first be examined followed by a short survey of the Iranian nuclear programme. After portraying the current status of research, the objectives of the present book will be formulated and the general structure outlined.

1.1 Nuclear Safeguards and Verification

The NPT for nuclear weapons was extended indefinitely in 1995. The five official nuclear powers, the USA, France, UK, China and Russia, and a further 182 non-nuclear-weapons states are parties to the treaty. The de facto nuclear weapons states India, Israel and Pakistan have not yet signed the treaty. The normative basis of the NPT is the twofold principle of support and surveillance. The nuclear weapons states are thus forbidden to pass nuclear weapons or nuclear explosive devices to other states or to assist them in developing nuclear weapons. Non-nuclear-weapon states, on the other hand, are obliged to give up any nuclear explosive devices and place themselves under international safeguards. In return, the non-nuclear-weapons

[3] IAEA Director General Statement to the General Conference of Member States. http://www.iaea.org/NewsCenter/News/2005/dg_gc49.html (November 2005).

states are granted free access to nuclear materials, facilities and technologies for the civil use of nuclear energy. Moreover, all contracting parties commit themselves to general and complete disarmament.

Two safeguards systems have developed on the basis of the NPT. In INFCIRC (Information Circular) /153 (IAEA/Hg. 1972), full-scope safeguards (FSS) are defined for non-nuclear weapons states. All the nuclear material of these states is subject to safeguards and monitoring by the IAEA . Furthermore, there are guidelines for states that have not yet acceded to the NPT although they have nuclear facilities or nuclear material. These safeguards are laid down in INFCIRC /66 (IAEA/Hg., 1968) and can be applied on the basis of commitments by the supplier countries. Voluntary safeguards are in place for the five official nuclear weapons states, USA, France, UK, China and Russia.

The IAEA's safeguards regime is based on *accountancy* and the principle of the *containment and surveillance* of nuclear materials. The essential objective of the IAEA's safeguards regime was and is the early detection of any diversion of significant quantities of declared nuclear material. The risk of detection by the IAEA is intended to deter member states from clandestine diversion. In the case of infringements of the treaty, the UN Security Council will be informed, who can decide on measures to be taken against the state on the basis of the UN Charter.

Since the IAEA's safeguards systems addressed the detection of the clandestine removal of declared nuclear material they were not in a position to detect any undeclared, clandestine nuclear programmes undertaken by the member states. Finally in 1992, the exposure of a secret nuclear programme in Iraq and North Korea led to a heated discussion on how to increase the efficiency of safeguards. In 1997, the IAEA's Board of Governors adopted the additional protocol (AP) INFCIRC/540 (IAEA/Hg., 1998) in order to be able to detect undeclared nuclear activities as well. This protocol is also known as the *Strengthened Safeguards System* and includes a large number of additional measures such as extended inspections, use of new technologies, procedures and methods, as well as the possibility of exploiting open sources of information (Fischer and Stein 1998).

For more detailed information on IAEA safeguards and the current discussion on this topic see also the texts of the safeguards systems IAEA/Hg. (1972, 1968, 1998) as well as Cooley (2006), Carlson (2006), Poucet (2006), Richard (2006) and Schenkel (2006).

The open information sources mentioned above do not only include maps and texts but also satellite images. The use of observation satellites has long been routine in disarmament control and within the framework of strengthened safeguards they can now also be used for verification purposes. The concept of verification will be defined in this work as in the field of disarmament control: a process for verifying compliance with contractual obligations. Verification is thus distinguished from monitoring by its specific, treaty-related character. According to Slink (1996), verification basically comprises three functions:

1. verification of compliance;
2. deterrence between the contracting parties;
3. building confidence between the contracting parties.

The process of verification involves the acquisition of data, the evaluation of their content and the subsequent reaction in the case of a treaty violation. Satellite-based remote sensing fulfils these three basic functions of verification and can help to support the verification process in the field of data acquisition and evaluation. It should, however, be emphasized here that the use of satellites can only be regarded as one of many tools for verifying the NPT. Satellite images can, for example, help in recognizing suspicious factors at an early stage and at low cost so that these factors can then be further examined by other means such as on-site inspections.

According to Kronfeld *et al.* (1993), the supplementary or preparatory tasks performed by satellite data are regular, large-area observations of the treaty area, or large sections of it, as well as specific observations of selected sites or objects. In this way, extensive activities and medium-term changes can be detected. This includes, for example, the construction of new nuclear facilities, changed use of civilian and military facilities, or the construction of new unlicensed objects in an existing complex. An overview estimating the potentials and challenges of commercial satellite data can be found in Jasani (2006).

Within the framework of the present book, a large-area observation with the aim of detecting (new) nuclear facilities will only be discussed briefly. This topic has already been dealt with thoroughly in Niemeyer (2001). Attention is rather focused on identifying objects in existing, known facilities and observing and interpreting changes over a prolonged period. To this end, use is made of multitemporal satellite images from the very high-resolution QuickBird 2[4] satellite, which has a spatial resolution of up to 0.60 m. This should make it possible to monitor the erection or modification of objects within a complex. This work picks up from studies in Niemeyer 2001 and investigates the potential of an object-based image analysis on the basis of very high-resolution satellite images.

1.2 The Iranian Nuclear Programme

Iran was one of the first states to sign the NPT (signature 1968, ratification 1970) and has thus committed itself not to undertake any activities in the nuclear weapons sector. In 1973, Iran signed the *Comprehensive Safeguards Agreement* (CSA) and in 1974 also the associated *Subsidiary Agreements* (SA). In February 2003, modifications to the Subsidiary Agreements were accepted and in December 2003 Iran signed the additional protocols to the Non-Proliferation Treaty. This means that full-scope safeguards are possible in Iran.

The nuclear infrastructure in Iran will be outlined in the following. The outline does not claim to be exhaustive and has an introductory function.[5]

[4] http://www.digitalglobe.com/ (September 2005).

[5] The information originates from the IAEA's Internet archive, http://www.iaea.org/ (November 2005), which contains a large number of documents concerning Iran. Further references

The Iranian nuclear programme began in 1959 with the purchase of a 5 megawatt (thermal) research reactor as well as a number of hot cells.[6] This reactor was erected at Teheran University and put into operation in 1967. In subsequent years, a large number of Iranian students were trained abroad (UK, USA, Germany and France) as nuclear physicists.

The reigning shah (Mohammad Reza Pahlavi) had ambitious plans for establishing a nuclear energy supply. Negotiations began with the USA, France and Germany for the construction of reactors. In 1975, a contract was signed with France for the construction of a nuclear research centre at Esfahan, followed by a contract with the German "Kraftwerk Union" for the construction of two 1300 MWe reactors at Bushehr. Another contract was concluded with France, and Iran also showed interest in uranium enrichment as well as the reprocessing of fuel rods.

The Islamic revolution of 1979 put a temporary stop to the expansion of the nuclear infrastructure. In the war with Iraq, 1980–1988, the half-finished Bushehr reactors were attacked several times and destroyed. From the mid-eighties onwards, nuclear activities were revived. China took over work on completing the research centre at Esfahan and in 1995 a contract was concluded with Russia, according to which a 1000 MWe reactor was to be built at Bushehr.

Some of the activities currently known to be in progress are listed in the following according to the individual facilities. Most of the facilities listed are also the subject of the investigations carried out in the present study. Table 1.1[7] once again summarizes all the Iranian facilities which are subject to IAEA safeguards together with their status. The individual facilities are described without any evaluation of the potential proliferation risk since opinions on this topic differ greatly. It is up to readers to form their own opinion.

Natanz. The Director General of the IAEA, Dr. ElBaradei, visited the plant in February 2003 after its existence had been made public in 2002. Nantanz includes a pilot plant for uranium enrichment consisting of about 1000 gas centrifuges and is currently in operation. According to official information, the facility will be used to produce low-enriched uranium (max. 5% U-235), which can be employed as fuel in research and light-water reactors. A large commercial facility with about 50,000 centrifuges is planned for the near future.[8]

Arak. A heavy water production plant, a 40 MWt research reactor and a number of hot cells are under construction. According to official information, the reactor moderated by heavy water serves for the production of radioisotopes for the medical and agricultural sector.

---

can also be found on the websites of http://www.globalsecurity.org, http://www.iranwatch.org, http://www.isis-online.org and http://www.fas.org.

[6] Small, heavily shielded rooms for handling radioactive materials.

[7] http://www.iaea.org/Publications/Documents/Board/2003/gov2003-75_annex2.pdf

[8] See for example: Nonproliferation Policy Education Center, "Iran: Breaking out Without Quite Breaking the Rules"; A Nonproliferation Policy Education Center Analysis.

Table 1.1 Iranian facilities subject to IAEA safeguards

Locations as of November 2003	Status
TEHRAN NUCLEAR RESEARCH CENTRE	
Tehran Research Reactor (TRR)	Operating
Molybdenum, Iodine and Xenon Radioisotope Production Facility (MIX Facility)	Constructed, but not operating
*Jabr Ibn Hayan Multipurpose Laboratories (JHL)	Operating
*Waste Handling Facility (WHF)	Operating
TEHRAN	
*Kalaye Electric Company	Dismantled pilot enrichment facility
BUSHEHR	
Bushehr Nuclear Power Plant (BNPP)	Under construction
ESFAHAN NUCLEAR TECHNOLOGY CENTRE	
Miniature Neutron Source Reactor (MNSR)	Operating
Light Water Sub-Critical Reactor (LWSCR)	Operating
Heavy Water Zero Power Reactor (HWSPR)	Operating
Fuel Fabrication Laboratory (FFL)	Operating
Uranium Chemistry Laboratory (UCL)	Closed down
Uranium Conversion Facility (UCF)	Under constr., first process units being commissioned for operation
Graphite Sub-Critical Reactor (GSCR)	Decommissioned
*Fuel Manufacturing Plant (FMP)	In detailed design stage, construction to begin in 2004
NATANZ	
*Pilot Fuel Enrichment Plant (PFEP)	Operating
*Fuel Enrichment Plant (FEP)	Under construction
KARAJ	
*Radioactive Waste Storage	Under construction, partially operating
LASHKAR AB'AD	
*Pilot Uranium Laser Enrichment Plant	Dismantled
ARAK	
*Iran Nuclear Research Reactor (IR-40)	In detailed design phase
*Hot cell facility for production of radioisotopes	In preliminary design stage
*Heavy Water Production Plant (HWPP)	Under construction. Not subject to Safeguards Agreement
ANARAK	
*Waste storage site	Waste to be transferred to JHL

*Facilities declared in 2003

Esfahan. The NFRPC (Nuclear Fuel Research and Production Centre) Esfahan is Iran's largest nuclear research centre. A plant for converting uranium oxide into uranium hexafluoride (UF_6) and metallic uranium will soon be completed. The production of UF_6 is a necessary step on the way to uranium enrichment. The nuclear technology center at Esfahan includes a reactor for producing neutrons (27 kW thermal), a heavy-water reactor (0 kW), a subcritical light-water reactor and a subcritical graphite reactor. A plant for the production of fuel rods for the heavy-water reactor in Arak is also planned for the near future.

Bushehr. A 1000 MWe pressurized water reactor of Russian design is under construction at Bushehr. Reprocessing of fuel rods in Iran itself has not yet been approved. After being used in Bushehr the fuel rods have to be returned to Russia.

Teheran. Teheran University is home to a 5 MWt research reactor as well as hot cells and reprocessing facilities on a laboratory scale.

Saghand. There are activities concerning mining uranium in the vicinity of Yazd.

A schematic overview of the sensitive technologies of the nuclear fuel cycle is given in Fig. 1.1. Two possible paths are shown for acquiring fissile material for atomic weapons. According to Wirz[9], there are civilian activities that could be indications of the fact that a country is attempting to use its nuclear industry to produce fissile materials for military purposes:

- uranium enrichment (uranium path);
- own production of fuel rods (plutonium path);
- reprocessing of spent fuel rods (plutonium path);
- construction of reactors that use heavy water or graphite to slow down the neutrons, and also facilities for the production of heavy water (plutonium path). Such reactors are especially sensitive because they use natural uranium, which, particularly if the country has its own uranium reserves, is much easier to procure than the low-enriched uranium usually employed in reactors.

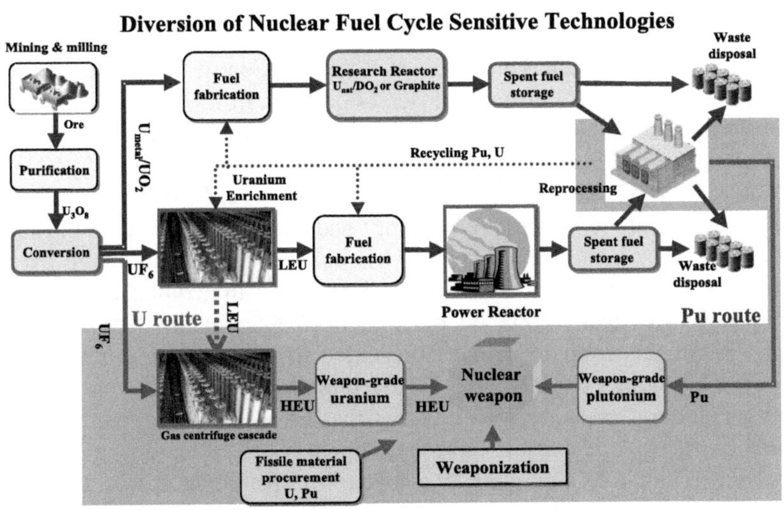

Fig. 1.1 Sensitive technologies in the nuclear fuel cycle. Modified after Richard (2006)

[9] Labor Spiez, Ist der Iran auf dem Weg zur Atombombe?, www.laborspiez.ch/d/h_info/iran/HI_Iran_Jan04_d2_int.pdf, (November 2005).

The question of whether Iran is on its way to building an atomic bomb cannot and will not be answered in this work. The main topic is rather identifying the potential of an image analysis of high-resolution satellite images for supporting verification of the NPT.

1.3 Current Status of Research

The present work is embedded in the support programme of the Federal Republic of Germany, represented by the Federal Ministry of Economics and Labour, reference no. 02 W6243, and Research Centre Jülich, Programme Group Systems Analysis and Technology Evaluation (STE). The results of the support programme "*Further Institutional and Technological Development of International Safeguards*" will also be made available to the IAEA.

On the European level, the topic of satellite-based verification of security-relevant topics is being dealt with by the GMOSS expert network . GMOSS[10] stands for *Global Monitoring for Security and Stability* and is implemented in the 6th EU Framework Programme. The mission of the GMOSS expert network is to integrate Europe's civil security research so as to acquire and nourish the autonomous knowledge and expertise base Europe needs if it is to develop and maintain an effective capacity for global monitoring using satellite earth observation.

GMOSS currently consists of 23 European partner organizations and 12 work packages. These work packages are divided into the main fields of applications, sociopolitical studies and generic technologies. The present study is incorporated into the *Treaty Monitoring* work package via the Programme Group Systems Analysis and Technology Evaluation (STE) at Research Centre Jülich. Other organizations with which there is currently close cooperation in the field of satellite-based nuclear verification are King's College London (Jasani, 2004), the Joint Research Centre in Ispra, the Freiberg University of Mining and Technology (Niemeyer and Canty, 2003; Niemeyer, 2005), and also the Federal Institute for Geosciences and Natural Resources in Hannover (Canty and Schlittenhardt, 2001). The use of object-based image analysis for this field is currently being discussed by STE at Research Centre Jülich and the Freiberg University of Mining and Technology (Niemeyer and Nussbaum,2005; Niemeyer *et al.*, 2005a, b; Nussbaum *et al.*, 2005). A large part of this research will be presented in this book. Figure 1.2 gives an overview of all partner organizations involved in GMOSS.

Another federation is represented by the ESARDA (*European Safeguards Research and Development Association*). This network of European organizations is involved in the R&D of nuclear safeguards (Avenhaus *et al.*, 2006). On an international level, a working group of the Canadian Atomic Energy Control Board is concerned with the potential of optical and radar image data (Truong *et al.*, 1999).

---

[10] http://gmoss.jrc.it/

Deutsches Zentrum für Luft- und Raumfahrt (Deutsches Fernerkundungsdatenzentrum DFD)	Germany
Definiens Imaging GmbH	Germany
Forschungszentrum Jülich GmbH	Germany
Bundesanstalt für Geowissenschaften und Rohstoffe	Germany
Universitet Linköpings Department of Computer and Information Science	Sweden
OD Science Application	Sweden
Swedish Defence Research Agency	Sweden
Commissariat à l'Energie Atomique	Frankreich
Centre national d'études spatiales	France
King's College London, Department of War Studies	UK
QinetiQ	UK
Centro di Ricerca Progetto San Marco	Italy
Università della Basilicata	Italy
Joanneum Research	Austria
University of Salzburg, Department of Geography and Geoinformation	Austria
Royal Military Academy Signal and Image Centre	Belgium
Technical University of Denmark	Denmark
The Netherlands Organisation for Applied Scientific Research	Netherlands
European Union Satellite Centre, Torrejon	EU
The Joint Research Centre	EU
UNOSAT	Schwitzerland
swisspeace	Schwitzerland

Fig. 1.2 Partner organizations in GMOSS

In the following, the present status of research will be discussed in more detail with respect to the applications and methods of an object-based approach. The internet portal of Definiens[11] gives a good overview of current work. Publications concerned with the object-based approach making use of eCognition software can be found there listed according to year of publication. This list makes no claim to be exhaustive. It can, however, be regarded as a focus for current work since at the moment there is no other commercial software available for object-based image analysis.

Object-based image analysis is applied in various fields such as the wide-area monitoring of landscape areas (Blaschke, 2005; Chandra *et al.*, 2005; Crase and Hempel, 2005; Laliberte *et al.*, 2005; Witheside, 2005); the monitoring of densely settled urban areas (Chunyang *et al.*, 2005; Grenzdoerfer, 2005; Moeller, 2005); for obtaining information in the case of disasters (Heremans *et al.*, 2005; Kouchi and Yamazaki, 2005; Bitelli *et al.*, 2004); or in data fusion and the establishment of GIS systems (Sim, 2005; Cabral *et al.*, 2005; Kosugi and Kosaka, 2005; Benz *et al.*, 2004; Langanke *et al.*, 2004).

These topics only represent examples since about 70 publications are listed for 2005 alone. It also becomes apparent that the field of methodological development is still in its infancy with respect to object-based image analysis. Most object-based work is concerned with the application of the existing eCognition software to different subject areas. At the moment, there are still very few new methods or algorithms which are capable of extending object-based work with eCognition. In the past few years, initial approaches have been made for example in the field of change detection (Blaschke 2005, 2004), image fusion (Chen *et al.*, 2005; Greiwe, 2004), segmentation (Frauman and Wolf 2005) and also building extraction (Liu *et al.*, 2005). Numerous discussions with other users of the software have revealed that additions and improvements are mainly required in the fields of segmentation and feature analysis. For both fields it would be useful to have a measure of quality which would systematize the work flow and ensure general comparability. There are some measures of quality for segmentation algorithms in general. For a comparative overview see Yang *et al.* (1995) or Zhang (1996). However, these measures have not been developed for the field of remote sensing and need to be adapted to this field in general and especially to eCognition software.

In the present study, an attempt will be made in Chapter 5 to improve the field of feature analysis. To this end, a feature analysis program has been developed, which at the same time also represents a measure of quality for features.

The object-based approach is still fairly new in remote sensing, but it is attracting more and more attention on the part of this community. "Googling"[12] leads to 171,000 hits for the term *eCognition*, in comparison to 430,000 hits for pixel-based software like *Geomatica*, 62,000 for *ENVI RSI* and 980,000 for *Erdas Imaging*. The GIS system *ArcGis* actually got as many as 5,550,000 hits. This comparison does

[11] http://www.definiens-imaging.com (July 2006).

[12] As of: March 2006.

not claim to be exhaustive, but in the author's opinion it does give some idea of the worldwide response.

1.4 Objectives and Structure of the Study

The present study analyses the potential offered by an object-based image analysis for supporting the verification of the treaty on the non-proliferation of nuclear weapons. The investigations are based on very high-resolution optical RS data from Iranian facilities. The objectives of this work also imply a modification as well as an innovation or further development, application and evaluation of working procedures, methods and algorithms with respect to the following specific issues.

- **Are new demands made of automated preprocessing of very high-resolution satellite images, and what methods and algorithms accommodate these requirements?** The RS data available are being continuously improved, which is reflected, for example, in improvements in spectral, radiometric and spatial resolution. In the present study, new methods implemented within the framework of project work at Research Centre Jülich and in cooperation with the GMOSS expert network will be presented, examined and evaluated with respect to their applicability with respect to very high-resolution RS data.
- **What methods and procedures are appropriate for (semi-) automated object-based image analysis in order to selectively obtain information on NPT verification from very high-resolution RS data?** In the present study, a new methodology will be developed, applied and evaluated. This methodology will comprise all steps within object-based image analysis such as preprocessing, segmentation, feature analysis, semantic modelling, classification and evaluation of results. Each step will be presented in detail, applied to RS sensing data for Iranian nuclear facilities and evaluated.
  In particular, the following points will be dealt with:

  - **Development of a new automated algorithm for the field of feature analysis. This will simplify and improve the semantic modelling and classification. Furthermore, it also provides a statistically derived measure of quality for image analysis of the features used.** The eCognition software provides a variety of features for semantic modelling and classification. In order to establish a rule-based classification model, the characteristic features have to be identified for each object class. eCognition does not as yet have a comprehensive statistical tool for this purpose, so that in most cases recourse must be taken to an empirical approach. The empirical approach is unsatisfactory in applying object-based image analysis as an automated process for information acquisition, nor does it make full use of the strength of a very broad feature basis.
  - **Extraction of typical features for the objects studied within nuclear facilities and establishment of reference models for object-based**

classification. Examination of the temporal and spatial transferability of such reference models. If typical features can be found for individual object classes or facility types then this means that the reference models may possibly be transferred to other RS data. If the results of this transferability are satisfactory then information from RS data can be provided rapidly and simply. This study will discuss and assess the creation of reference models and their spatial and temporal transferability.
- **The advantages of an object-based methodology in working with very high-resolution optical RS data will be clearly demonstrated. In particular, the quality of the classification results will be evaluated quantitatively.** The use of the more recent object-based approach only makes sense if it provides advantages in comparison to other methods. The advantages of this approach will be presented and the quality of the results of this new methodology discussed.
- **The suitability of the eCognition software for the questions to be studied will be examined thoroughly.** To this end the mode of operation of the software will be investigated in detail. Attention will be especially paid to its innovative approach and potential but also to future improvements and extensions.

- **How can multitemporal modifications to individual objects within a (nuclear) facility be automatically recognized and semantically defined?** The pixel-based change detection technology of *Multivariate Alteration Detection* (MAD) is very well suited for detecting significant changes in satellite images. The determination of change, that is to say the question: "What has changed?" is, however, only possible to a very limited extent with MAD technology. In this study, a new methodology has been developed for the detection **and** identification of multitemporal changes in very high-resolution RS data. A combination of pixel-based alteration detection and object-based image analysis will be presented for the first time. It will be shown that this combination provides decisive advantages and a gain in information in comparison to an uncombined procedure. The applicability of the MAD technique to very high-resolution RS data will also be examined and assessed.
- **What is the potential of an object-based methodology for supporting verification of the NPT with the example of Iranian facilities?** Since as a rule only a few sources provide information on the clandestine nuclear activities of individual states, attention will be focused on the way in which verification-relevant information can be obtained from satellite-based RS data for Iran. Furthermore, it will be shown which objects within (nuclear) complexes can be identified in very high-resolution optical satellite data. The issue of the long-term monitoring potential of satellite-based NPT verification will also be examined.

The present study is divided into 9 chapters, dealing with basic principles in Part I and object-based analysis and change detection of nuclear facilities in Part II.

After the introduction in Chapter 1, the basic principles will first be outlined in Part I. Chapter 2 gives a general overview of high-resolution RS data (Section 2.1) as well as of the two general methods of RS data analysis, which take radically different approaches. In Section 2.2, the well-known and frequently described methodology of pixel-based remote sensing will only be touched upon, whereas the more recent method of object-based image analysis will be discussed in greater detail in Section 2.3. In Chapter 3, the function and operating mode of the eCognition software will be considered more closely. The basic steps involved in object-based image analysis, i.e. (multiscale) segmentation (Section 3.2), semantic modelling (Section 3.3) and classification (Section 3.4), will be dealt with here. The final chapter on basic principles (Chapter 4) is concerned with change detection. The basic techniques of change detection are presented in an introductory chapter (Section 4.1). In Section 4.2 a new pixel-based method for change detection, multivariate alteration detection (MAD), will be described in detail and derived.

Part II is concerned with the derivation, application and evaluation of the new methods developed in this study for object-based image analysis and change detection against the background of the objectives formulated above. Before starting the actual image analysis, a new feature analysis tool SEaTH will be developed in Chapter 5. The derivation of SEaTH is the subject of Section 5.1, while Section 5.2 validates the new SEaTH methodology.

Chapter 6 is concerned with the application and evaluation of preprocessing the RS data studied. The geometrical correction of the RS data is one of the most important preconditions for change detection. A routine drawn up within the framework of the GMOSS network that efficiently automates this step for high-resolution data is discussed and applied in Section 6.1. Section 6.2 is concerned with data fusion, especially panchromatic sharpening. Four recent methods (Section 6.2.1–6.2.4) are derived and also qualitatively and quantitatively evaluated. Two of these methods have been implemented in the course of project work at Research Centre Jülich. As part of the transferability tests for reference models formulated in the objectives a methodology developed in the project work for automatic radiometric normalization will also be applied and assessed.

The methodology developed for the object-based analysis of nuclear facilities in Iran will be derived, discussed and evaluated in detail in Chapter 7. The methods developed in the present study will be applied as well as those from project work at Research Centre Jülich and GMOSS. Section 7.1 describes the application and evaluation of object-based analysis methods for RS data from the Nuclear Fuel Research and Production Centre (NFRPC) Esfahan 2002. Studies on the temporal transferability of the classification model established in the previous chapter will be performed in Section 7.2 for RS data from NFRPC Esfahan 2003 and 2004. Section 7.3 deals with the possible spatial transferability of an existing reference model. The application of the SEaTH method to low-resolution RS data for detecting undeclared (industrial) facilities with the example of various sites in Iran is the subject of Section 7.4. The potential analysis for the Bushehr site in Section 7.5 shows the basic opportunities offered by object-based analysis, without being restricted by software boundaries.

Chapter 8 is concerned with investigations of multitemporal change detection with the example of the NFRPC Esfahan site. To this end, a new methodology of combined change detection is derived, applied and assessed. In Section 8.1 the results of pixel-based change detection (Section 8.1.1), object-based change classification (Section 8.1.2) and class-related change determination (Section 8.1.3) will be derived in detail for the period from 2002 to 2003 and applied. These steps are summarized in Section 8.2 for the period from 2003 to 2004.

In Chapter 9 the results of the individual investigations are summarized and briefly discussed once again. The study ends with a conclusion.

Part I
Basic Principles

Chapter 2
Satellite Imagery and Methods of Remote Sensing

2.1 High-Resolution Satellite Imagery

Remote sensing (RS) interprets electromagnetic energy that is emitted and reflected by any object whose temperature is not at absolute zero. The energy propagates in the form of a *wave corpuscular dichotomy* (Mather, 2004). Satellite systems are either active or passive. Passive systems directly or indirectly use solar radiation, whereas active systems (such as radar systems) generate electromagnetic radiation themselves.

This study exclusively works with the RS data from passive systems. Solar radiation can be perceived by the human eye in the wavelength interval from 0.38 μm (blue) to 0.71 μm (red).[1] Wavelengths beyond this are in the infrared region and not visible to the human eye.

The basic principles of remote sensing have already been discussed and explained in the specialist literature on numerous occasions. In the following, a brief and concise introduction will be given, but a comprehensive explanation of the fundamentals will not be given in the present study; instead, reference will be made to the relevant literature, such as Campbell (1996), Jensen (2004), Mather (2004) or Schowengerdt (1997).

2.1.1 Spectral, Radiometric and Spatial Resolution

The **spectral resolution** of remote sensing systems is understood to be the wavelength intervals which the sensor of the satellite employed can detect. Existing remote sensing systems have different spectral resolutions and therefore also provide different image data. An overview of the spectral resolution of commonly used satellites is given in Fig. 2.1. For a detailed description of the radiation-physical fundamentals, see Kraus and Schneider (2001) or Schott (1997).

---

[1] The interval limits vary in the literature. However, they are always very close to the specified range.

S. Nussbaum, G. Menz, *Object-Based Image Analysis and Treaty Verification*,
© Springer Science+Business Media B.V. 2008

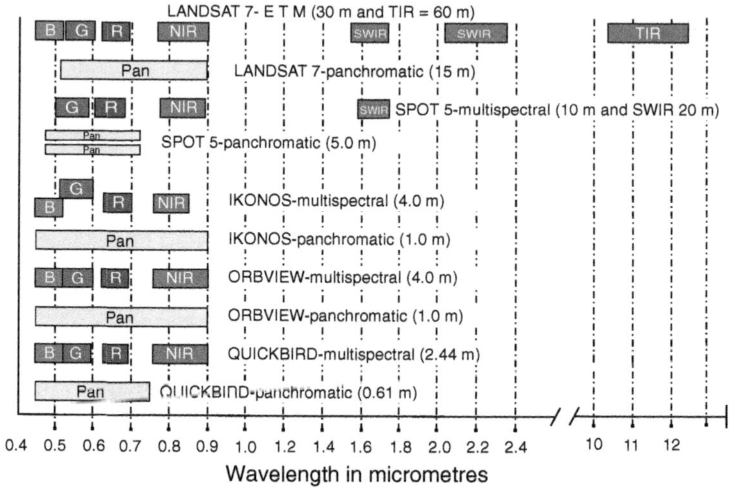

Fig. 2.1 Comparison of the spectral and spatial resolution of different satellite systems

The **radiometric resolution** of a satellite is a measure of the lowest discernible electromagnetic radiation. Depending on the detector used, this varies between 6 bits (64 grey values) and 11 bits (2048 grey values) for modern systems such as IKONOS or Quickbird-2.

Apart from spectral resolution, satellite systems also differ in their **spatial resolution**. The most frequently used form of measuring this resolution for satellites is the *instantaneous field of view* (IFOV). IFOV is defined as the area directly observed by a sensor. A satellite detector consists of sensors. Each of these sensors records a certain area represented by a pixel. The total image is composed of these pixels. The spatial resolution of an image depends on the real area mapped by an image pixel. Figure 2.1 shows the spatial resolutions of some satellite systems in metres. The higher or better the resolution of a detector, the smaller are the objects that can be identified on the satellite images.

In this study, very high-resolution satellite images are used. This is taken to mean a spatial resolution of 1 m or better in the panchromatic band and a corresponding resolution in the multispectral bands (cf. Fig. 2.1). The following commercial global satellites can be taken into consideration:

- IKONOS (panchromatic 1 m, multispectral 4 m)
- OrbView-3 (panchromatic 1 m, multispectral 4 m)
- Quickbird-2 (panchromatic 0.61 m, multispectral 2.44 m)

For further information such as technical data or orbital parameters, see the relevant satellite manuals or detailed documentation on the Internet.[2]

[2] IKONOS: http://www.spaceimaging.com, OrbView: http://www.orbimage.com and Quickbird-2://www.digitalglobe.com

2.1.2 Representation of Image Data

The image data in remote sensing are represented as image matrices consisting of $i = 1, \ldots, I$ image lines and $j = 1, \ldots, J$ image columns. A pixel at position (i, j) can be defined, according to its reflection values in all N bands of the satellite, as a pixel vector

$$\mathbf{x}_{ij} \tag{2.1}$$

This vector $\mathbf{x}_{ij}$ represents the grey values (0,1, ..., 255 for 8-bit data or 0,1, ..., 2048 for 11-bit data) of the pixels in the individual bands. Therefore 0 = black (meaning no reflection, maximum absorption) and 255 or 2048 = white (meaning maximum reflection, no absorption). Figure 2.2 schematically shows the vectorial

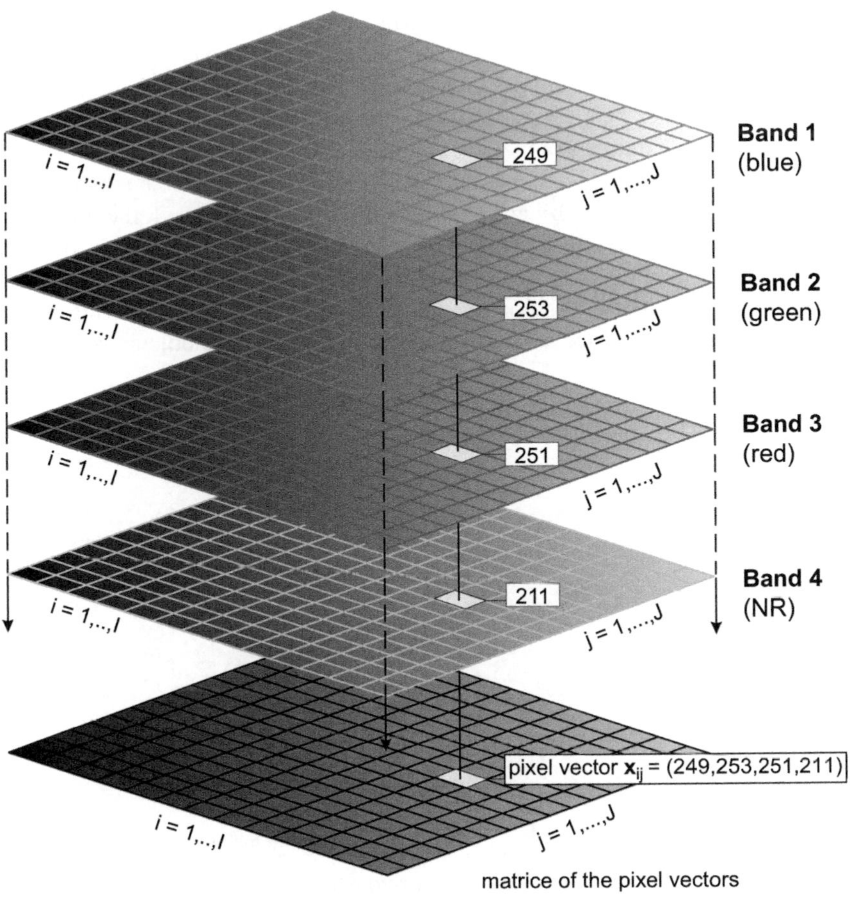

Fig. 2.2 Vectorial nature of the image pixels of a multispectral satellite scene (4 bands)

nature of the image pixels with the example of a multispectral scene. The individual satellite scenes used in this study are described in more detail in the respective investigations.

2.1.3 Spectral Signature

A basic assumption in remote sensing is that specific surfaces such as rocks, water or vegetation have characteristic reflection or emission behavior and can be distinguished by the specific **spectral signature**. This signature is influenced by many factors such as the material properties of the surface, the topography or the modification of radiation by the atmosphere (Richards and Jia, 1999). Figure 2.3 shows the spectral signature of selected surfaces. For reasons of clarity, typical signatures are plotted. It should be noted here that the signature changes depending on the condition of the surface considered. The reflectance of water, for example, depends on whether the body of water is clear or turbid, calm or agitated.

With the aid of these typical reflectances, the surface observed by satellite sensors can be determined in more detail. As a rule, the surface is either determined manually or by computer making use of a classification scheme. The satellite data are divided into a finite number of classes of identical significance in terms of content.

For the classification of satellite data, two basic methods are to be distinguished: the **pixel-based** and the **object-based** method. In the case of the pixel-based method, each individual image pixel is analysed and classified according to its spectral features (cf. Section 2.2). Object-based methods are based on the assumption that a pixel is very likely to belong to the same class as its neighbouring pixel. In a first step, the image space is segmented into homogeneous objects consisting of

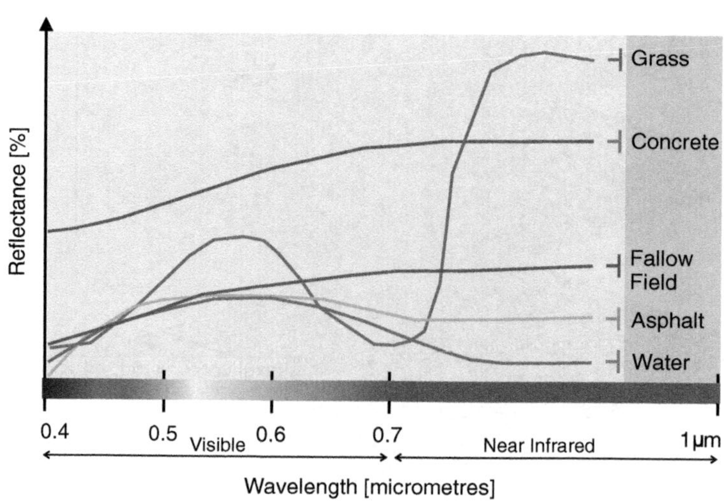

Fig. 2.3 Spectral signature of different surfaces, modified after Richards and Jia (1999)

similar pixels. These objects are then also grouped in classes of the same semantic significance. For classification, however, apart from the spectral features, additional features such as shape, size, texture and neighbourhood relations of the objects are available (cf. Section 2.3) (Koch *et al.*, 2003).

2.2 Pixel-Based Image Analysis

Most pixel-based classification methods can be subsumed under the term cluster analysis. In this case, the individual N-dimensional pixel vectors in an N-dimensional feature space are analysed, for example, with the aid of statistical methods, with fuzzy-logic techniques or with neural networks and assigned to a class. The feature to be classified is generally the spectral signature of a pixel. The aim of all these methods is the external separation of internally homogeneous classes, i.e. all (spectral) values allocated to a class should be as similar as possible, the differences of the (spectral) values of different classes should be as large as possible. Classification can be unsupervised or supervised with the aid of reference data and training regions.

Pixel-based methods are hitherto the most commonly used type of classification in remote sensing. Due to this fact, they are described in detail and mathematically derived in the specialist literature. The focus of this study is on the object-based method, so that for a precise explanation of the individual pixel-based methods reference is made to the scientific literature such as Albertz (2001), Canty (1999), Lillesand *et al.* (2003) and Richards and Jia (1999).

In recent years, these purely pixel-based methods have increasingly reached their limits despite further developments. One reason is the fact that with increasing spatial resolution of the satellite data the (small) feature basis of spectral values often only provides insufficient results for classification. Furthermore, there is an increasing amount of additional data such as information from GIS systems or digital elevation models (DEMs). It appears meaningful for future investigations to integrate these additional data into satellite image analysis. Object-based image analysis offers great potential here, since it has a very large feature basis for classification and additional data from other data sources can be readily integrated and used for analysis.

2.3 Object-Based Image Analysis

This chapter introduces briefly the object-based image analysis. Therefore the parts segmentation, sematic modelling and classification are discussed in detail. A special emphasizes is given to the segmentation and its requirements, performance and reproducibility.

Computer-controlled, object-based image analysis can be compared in a first approximation with visual perception. Thus, a viewer in visual image interpretation consciously recognizes specific shapes and correlations that go beyond the pure grey values and color grades of the image. He perceives typical patterns and associates them with real objects. Apart from pure color information, these patterns result from other features such as texture, shape, size or from the relations between individual objects. A procedure similar to visual interpretation is also aimed at by object-based image analysis in general and the eCognition software used for image analysis in particular. Although the complexity and performance of human perception is by no means reached, there are certain parallels (cf. Chapter 3).

Object-based, knowledge-based image analysis methods rank among the so-called computer-vision techniques (Wilkinson, 1999). They have hardly played a role in remote sensing so far. Due to growing demands made on remote sensing and its increasing possibilities, the functioning of the usual pixel-based methods must be reassessed and redesigned. On account of the large potential of object based image analysis, recent development work increasingly goes in this direction, as does the eCognition software from Definiens used in this study.

An image to be analysed is segmented into individual image objects. The segments formed should well reflect the objects present in reality. Various methods can be used for segmenting. The objects can be classified by heuristic methods or by knowledge-based techniques such as semantic networks. In knowledge-based systems, the rule base for classification is established via a feature combination. Apart from the spectral signature, features such as shape, size, texture and neighbourhood relations of the objects can additionally be used for object description. A rating function is then used to allocate the objects to a number of classes according to their feature description.

Knowledge-based, object-based image analysis integrates knowledge about the objects into the analysis. A differentiation is made here between declared knowledge (knowledge from various data sets), procedural knowledge (method for utilizing the data sets) and a priori knowledge (knowledge without relation to the analysing data inventory). The investigations and analyses carried out in this study were performed against the background of a knowledge-based, object-based image analysis, where all types of knowledge available about the study region are integrated into the process.

2.3.1 Segmentation – Requirements, Performance and Reproducibility

The object-based methodology segments the image into individual objects that are homogeneous in themselves and form the basis for further image analysis. This process – termed segmentation – has already been used for some time in pattern recognition and materials monitoring and numerous methods are available (Blaschke 2000). A good overview is given by Haralick and Shapiro (1992), Sagerer and Niemann (1997) as well as Sonka *et al.* (1998). An extensive listing of basic approaches with associated current work can be found in Niemeyer (2001).

Put simply, there are three main criteria according to which the different segmentation algorithms combine the individual image pixels to form objects: the homogeneity within a segment, the separation from adjacent segments and shape homogeneity (simplicity of segment boundaries vs. fractal attribute). These criteria cannot all be fully met simultaneously, since they in part exclude each other. Many segmentation methods therefore particularly emphasize certain criteria.

A detailed explanation of all the methods is beyond the scope of the present study and reference is made to the above-mentioned studies. The multiscale segmentation algorithm of the eCognition software used in this study is derived and described in detail in Section 3.2.

In principle, there exist so-called edge-based algorithms and region-based algorithms for the different methods. Edge-based methods start segmentation with a search for strong gradients in the image. It is assumed that these discontinuities constitute segment boundaries. The edges searched for by edge detectors are built up to form a network of boundaries. Especially the criterion of the (simple) shape of segment boundaries can be taken into consideration here. Region-based algorithms, starting from an initial region, decide according to predefined criteria whether or not a particular pixel is allocated to this region. In the top-down approach, the total image is segmented into ever smaller objects until all the segments satisfy the predefined criteria. The bottom-up approach is based on so-called seed pixels. These are fused with neighbouring pixels to form a segment, until certain truncation criteria are exceeded.

2.3.2 Requirements for Segmentation

The numerous segmentation methods already developed for other areas such as medicine, neuroinformatics or telecommunications are very specialized and can hardly be transferred to other fields such as remote sensing. According to Schiewe and Tufte 2001 the main reasons for this are

- that in remote sensing, multispectral and in part also multiscale input data arise, so that not only the complexity but also the redundancy and thus the reliability of the methods clearly increase;
- that a large variety of additional data (e.g. information from GIS systems) is available and can be integrated into the process;
- that, in comparison to other applications, various objects with heterogeneous properties (with respect to shape, spectral behaviour, textures, etc.) have to be classified and that an integration of these different properties is not yet adequately taken into consideration;
- that especially objects of different sizes are viewed and existing methods do not permit adequate multiscale applications.

As a consequence of these requirements, software for the object-based treatment of RS data must be able to treat different data sources simultaneously, it must be possible to apply several segmentation strategies, depending on the objective, and different scale levels of segmentation must exist in parallel, since heterogeneous objects of different sizes occur in a satellite image. Schiewe (2002) terms these requirements the *multi-source*, *multi-method* and *multi-scale aspects*.

In the field of remote sensing, the first commercial object-based image analysis software, eCognition from Definiens Imaging, has been on the market since 2000. In this software, a multiresolution segmentation algorithm is implemented. Furthermore, there are various segmentation modules on the market which can be integrated into existing remote sensing programs. In the following, the performance of current segmentation algorithms will first be briefly evaluated and then it will be shown whether and how the above requirements for a segmentation procedure are fulfilled.

2.3.3 Algorithm Performance

A comparison of different segmentation programs is made by Neubert and Meinel (2003). They first generally compare a selection of currently available segmentation algorithms and then determine the quality of the segmentation results on the basis of very high-resolution, panchromatically sharpened IKONOS multispectral data. The following programs were selected for the comparison:

- eCognition 2.1 bzw. 3.0 (Definiens Imaging);
- Data Dissection Tools (INCITE, Stirling University);
- CAESAR 3.1 (N.A. Software);
- InfoPACK (InfoSar);
- Image Segmentation für Erdas Imagine (Remote Sensing Application Center);
- Minimum Entropy Approach to Adaptive Image Polygonization (Rheinische Friedrich-Wilhelms Universitdt Bonn, Institut für Informatik).

Segmentations were generated using two test regions. The aim was to extract relevant object boundaries of land cover and land use. Good segmentation is achieved if as many heterogeneous areas as possible are correctly demarcated and, on the other hand, homogeneous areas are not excessively segmented into subsegments (oversegmentation). Moreover, the shape criteria determined should conform as closely as possible to the dimensions of the reference areas.

In their comparison, Neubert and Meinel came to the conclusion that the eCognition software provides good results and offers great potential due to multiscale segmentation and the possibility of image classification based on fuzzy logic. They emphasize the reproducibility of segmentation and the diversity of interfaces with other GIS and remote sensing systems. Apart from the good results, however, irregularly delimited segments are also formed in some cases. Good results were also obtained with InfoPACK, although it does have a tendency towards oversegmentation and the formation of tubular segments at the transitions. The results of the other approaches do not come up to these two programs (Neubert and Meinel 2003).

2.3.4 Reproducibility of Results

An important and decisive aspect of a high-performance segmentation algorithm is the reproducibility of its results. The result of a segmentation must be reproducible at any time. This type of reproducibility is 100% given for the eCognition software used, as already mentioned by Neubert and Meinel.

Conceivable, however, is yet another type of reproducibility – called *independent reproducibility* in the following. Here, an object that has remained unchanged in two satellite scenes at different dates should be identically segmented and identified as the same object. This would permit a realistic change detection on the basis of an object-based classification. However, this independent reproducibility is not yet given at present. The reason is, on the one hand, the dependence of segmentation on various parameters in recording the image. These parameters in part vary significantly in differently viewed periods of time. Examples of changing parameters are: radiometry, illumination or the different camera angle. They lead to a different segmentation of identical objects in individual multitemporal scenes.

Another reason for the lack of independent reproducibility is the operating mode of the multiscale segmentation used (cf. Section 3.2), which determines its starting points on the basis of all the image information available. Thus, the segmentation of a satellite scene and that of a subset from this scene leads to different results for the object boundaries. Since the parent population in the comparison of a scene and a scene segment is different, different starting points are chosen. This may lead to a different segmentation, for example, of one and the same building. This results in artificial changes caused by the segmentation, which reduces the quality of the result of a change detection on the basis of a pure classification comparison.

Independent segmentation in object-based image analysis is at present a rather utopian demand, since the problem of changed recording conditions in multitemporal images is already familiar from pixel-based classification methods. Against the background of change detection and the further investigations in the present study, however, this should be borne in mind as a given restriction. Since the first step in object-based change detection is the segmentation, all further results obtained are based on segmentation performance.

A partial solution to this general problem of change detection is an appropriate pre-processing of the satellite data, where the individual scenes are matched as optimally as possible in various steps. As a rule, this step precedes all further measures. The methods applied may vary (cf. Chapter 6).

Apart from a simple classification comparison for change detection, other techniques can also be applied (cf. Chapter 4). Finally, a combination of pixel-based change detection and object-based classification should be taken into consideration (cf. Chapter 8), since the problem of differently segmented object boundaries does not arise in pixel-based change detection. A new type of cluster or segmentation algorithm would also be conceivable. This would have to pay more attention to other features such as shape features or position of the objects in the image and be able to selectively extract objects in the image. Since the development of such an algorithm would go beyond the scope of this study, the other above-mentioned possibilities are applied and the existing multiscale segmentation algorithm of the eCognition software is used

The eCognition software developed by Definiens Imaging in 1999 and continuously extended since then fulfils important demands for a product to be used in remote sensing:

- multiscale segmentation that can be influenced by the user;
- the possibility of incorporating additional widely differing data;
- an integration of different heterogeneous features for classification.

This fulfils the multi-source, multi-method and multi-scale aspects required by Schiewe (2002). It may therefore be assumed that the segmentation algorithm used is one of the currently most efficient algorithms for remote sensing.

2.3.5 Semantic Modelling and Classification

An object can be described by characteristic features such as its spectral signature, shape, size, texture and neighbourhood relations to other objects. These representative features enable a later classification into object classes. For semantic modelling, these typical features of the object classes must first be determined and combined in a semantic model. In this study, the representative features are determined using the feature analysis program SEaTH (cf. Chapter 5).

The model is constructed as a rule-based semantic network, which constitutes a form of knowledge presentation. The advantage of a network is the possibility of a hierarchical arrangement and the interlinkage of all information on the objects. The individual object classes in the model are termed nodes. Concepts provide the generic description of the object classes. The implementation of these concepts in the satellite scene under consideration is called instances (Tönjes, 1999; Quint, 1998).

The nodes of the network are linked to each other by edges. These edges describe the relations between neighbouring, lower or upper objects. In technical terms, these relations can be broken down into hierarchical, topological, optional and obligatory relations. Hierarchical and topological relations result from the structural relations between the objects, whereas optional and obligatory relations are determined via the minimum and maximum number in the relations of the network.

The features of an object class are described by attributes. They have a measured value and a value range to which methods for calculation are allocated. The knowledge inherent in a semantic network can be separated into declared and procedural knowledge. Declared knowledge contains the concepts and relations, whereas procedural knowledge comprises the methods for calculating the concept attributes and for evaluating the concepts and relations. The procedural knowledge thus determines the order of the analysis, which can be presented by a set of rules.

This rule base contains a condition and an action part. The condition verifies whether there is a new interpretation state for the neighbouring node in the semantic network and the action part adapts this state accordingly. If there are several conditions characterizing an object class, they are linked by (logical) operators [e.g. and, or, not].

For a detailed description of semantic networks, see the extensive literature available on this topic such as Förstner *et al.* (1999), Sagerer and Niemann (1997), Tönjes (1999) and Tönjes *et al.* (1999).

The semantic models of this study are set up as a network of object classes on the basis of a rule base, as described above. The object classes are described by representative features.

The semantic classification determines whether or not an object belongs to a certain object class on the basis of its significance. A rating function determines a level of confidence by comparing the individual features of the object to be classified with the attributes. This confidence level provides information about whether the object belongs a class, so that the rating function is also called the membership function. The membership can be determined by various approaches. Widely used approaches are Bayesian probability models (Centeno, 1998; Haralick and Shapiro, 1993), the evidence theory of Dempster and Shafer (Centeno, 1998; Kunz *et al.*, 1998) or the models based on fuzzy logic (Baatz and Schäpe, 1999; 2000; Centeno, 1998). The classification methods used in this study are discussed in Section 3.4.

Chapter 3
eCognition Image Analysis Software

eCognition is a software product from the Munich Definiens company, which was founded by the Nobel laureate Gerd Binning. Object-based image analysis software is in principle suitable for use in all fields of remote sensing. The software has been continuously developed since 1999 so that any weak points that may have been detected have contributed to a further improvement of the software. Since all the object-based analyses in the present study were performed with eCognition, the basic principles, modes of operation and algorithms of the software will be described and derived in detail in the following. In this study, eCognition is used against the background of object-based image analysis and change detection on the basis of high-resolution satellite images. The criticisms, suggestions for modifications or extensions put forward in the individual chapters are therefore not of a general nature but relate specifically to this background.

In spite of possibilities for improvement and some proposals for extending the product, which, however, apply to any software that is being further developed, eCognition was selected for this project. It is the only commercial object-based analysis software specially developed for the field of remote sensing. It offers new potential and fulfils decisive criteria including the important multisource, multimethod and multiscale aspects (see Sections 2.3 and 3.2). Furthermore, eCognition is characterized by the powerful segmentation algorithms currently available and the possibility of hierarchical, cross-linked modelling and classification.

3.1 Structure and Workflow

The basic idea behind eCognition is that image objects and the relations between these objects contain important semantic image information. The image analysis software copies the way in which human perception functions. The human brain constructs an overall image from individual objects. The objects are related on the basis of a large number of features with empirical values, existing knowledge and the surrounding objects. The eCognition software works in a similar but greatly simplified manner, and is by no means as complex. In the first step, the image

S. Nussbaum, G. Menz, *Object-Based Image Analysis and Treaty Verification*,

© Springer Science+Business Media B.V. 2008

pixels from the image are grouped to form objects with the aid of a multiscale segmentation (cf. Sections 2.3 and 3.2). The simultaneous formation of several levels of segmentation on different scales is possible and indeed desirable. These levels are then hierarchically linked to each other. Since image objects, image information and processes are present simultaneously on several scales and the relation between these scales is known, important additional information can be obtained for interpretation purposes and real-world objects of different sizes can be classified.

After multiscale segmentation, the features and relations of the individual object classes are defined within a hierarchical network. This step is also termed semantic modelling (cf. Section 2.3, 3.3). The semantic classification of the objects into object classes is then performed on the basis of this semantic model (cf. Sections 2.3 and 3.4). Figure 3.1 gives a schematic overview of the general workflow with eCognition.

A drawback of eCognition is the lack of capability for the necessary preprocessing of the RS data (see Chapter 6). These preprocessing steps are indispensable for large areas of remote sensing and in particular for the field of change detection. This means that recourse must always be taken to an additional remote sensing program and it is therefore hardly possible to work with eCognition alone in the field of remote sensing. It must also be mentioned that there is no possibility of independent reproducibility, as already mentioned in Section 2.3. Furthermore, there is no comprehensive, statistical or really satisfactory tool for the field of feature extraction. It is very tedious and time-consuming to filter out those features and threshold values that describe an object class well from the large number of possible (predefined) features and values.

Since eCognition is the only commercial software of its type and it continues to be further developed and improved, the criticisms expressed and the programs developed can only help to improve the software. Nevertheless, eCognition is an innovative software which has great potential and enables new working methods to be employed in the field of object-based image analysis.

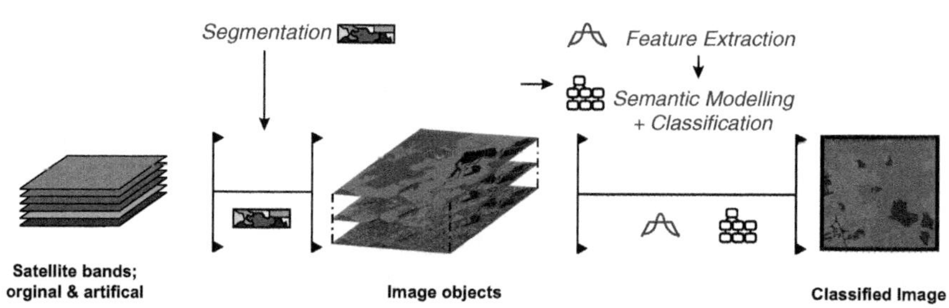

Fig. 3.1 Workflow of object-based image analysis in eCognition

3.2 Multiscale Segmentation

The multiscale segmentation used in the eCognition image analysis system fulfils important demands made on a segmentation method for the remote sensing field. These requirements are the multisource, multimethod and multiscale aspects already mentioned in Section 2.3. In comparison to segmentation methods currently in use, the eCognition image analysis software can be regarded as one of the most powerful systems available at present for the field of remote sensing (cf. Section 2.3). The segmentation procedure permits an extraction of various sizes of image objects in a project (multiscale aspect) so that real-world objects of any magnitude can be identified and classified. Furthermore, it permits the segmentation strategy to be modified depending on the goal. This can be varied by weighting the parameters of *color* and *shape*, and the choice of the neighborhood (multimethod aspect). The multisource aspect is fulfilled by the possibility of incorporating various types of additional information (e.g. GIS data, indices or DEMs) (cf. Fig. 3.2).

Other demands made on an efficient algorithm are the reproducibility of the results and an acceptable speed, even for large data sets. Reproducibility in the sense of repeating the segmentation of the same satellite scene that has already been segmented is completely fulfilled by eCognition. Other aspects and questions of reproducibility have already been mentioned and discussed in Section 2.3. The speed can be described as acceptable even if the segmentation of large high-resolution satellite scenes takes a certain amount of time. In the 4.0 version of eCognition the problem of reduced speed with large data sets has been tackled by a large data handler.

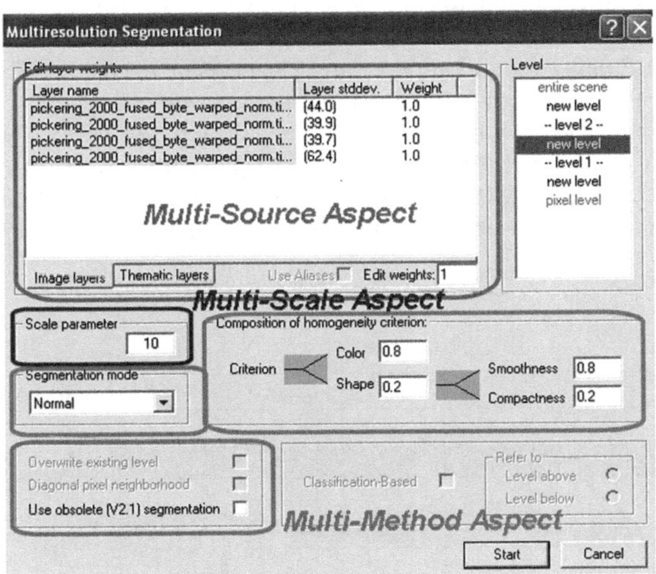

Fig. 3.2 Dialogue box for multiscale segmentation (eCognition Version 3.0). Compliance with the multisource, multimethod and multiscale aspects is shown in different colors

The aim of segmentation is the construction of a network of image objects in which smaller image objects are subobjects of larger superordinate objects. This hierarchical network simultaneously represents information from image objects of various sizes. It is thus also possible to process very heterogeneous images, such as high-resolution satellite scenes or aerial photographs, that have real-world objects on various scales (e.g. buildings, districts of a town, and towns and cities in a landscape). By means of the network structure each image object "knows" its context, its neighborhood as well as its superordinate and subordinate objects, and relations can be defined between these objects. Figure 3.3 gives a schematic representation of the principle of the hierarchical network of individual segmentation levels, and Fig. 3.4 illustrates this with real RS data. This network of image objects forms the basis for the object-based classification in eCognition.

The method of multiscale segmentation is based on the assumption of a color and spatial continuity of individual image objects. The segmentation algorithm is a heuristic optimization procedure which minimizes the mean heterogeneity of the image objects for a given resolution over the entire scene. A minimization of the heterogeneity leads to more homogeneity so that in the following the expression maximization of homogeneity will be used. In order to optimize the homogeneity, the mean homogeneity of the image objects together with their size (number of pixels) is weighted, and the sum of this weighted homogeneity is maximized over all the image objects, see Baatz and Schäpe (2000).

Multiscale segmentation functions as a region growing method. This is regarded as a region-based algorithm (cf. Section 2.3). Here, the eCognition image analysis system follows the bottom-up approach. Starting from individual pixels, the segments grow on the basis of a large number of pairwise data clusterings (cf. Hofmann *et al.* (1996) and Hofmann and Buhmann (1997)), which take place regularly and uniformly over the entire picture. The decision to coalesce is taken on the basis of local *homogeneity criteria* and *optimization procedures*, which maximize the mean homogeneity of the image objects and the reproducibility. The discussion of the homogeneity criteria and the optimization procedures follow the descriptions in Baatz and Schäpe (2000). Here, use is made of the concept of maximization of homogeneity, whereas Baatz and Schäpe speak of a minimization of heterogeneity.

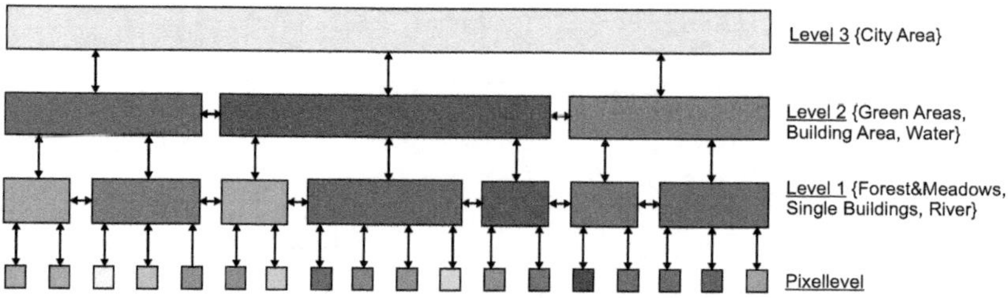

Fig. 3.3 Structure of a hierarchical network. Braces indicate possible, fictitious semantics

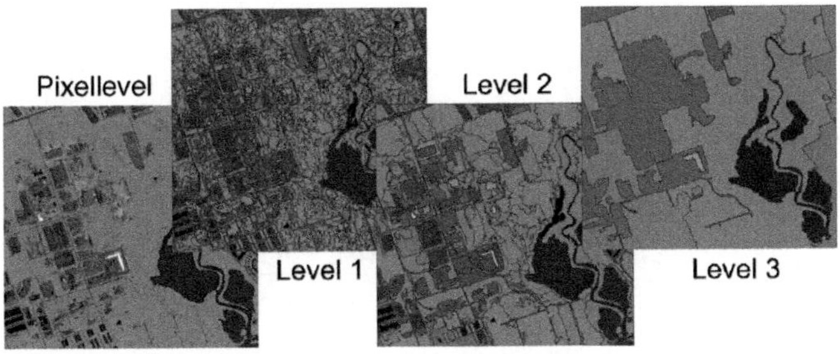

Fig. 3.4 A hierarchical network of segmentation levels in eCognition

3.2.1 Homogeneity Criteria

In a given *d-dimensional* feature space, two adjacent image objects (or initially image pixels) f_{1d} and f_{2d} are termed similar if their features in the feature space are closely spaced; i.e. they are at a short distance[1] from each other. The shorter the distance the more likely they are to belong to the same object. The *degree of fitting* results from:

$$h = \sqrt{\sum_d (f_{1d} - f_{2d})^2} \tag{3.1}$$

This difference can be normalized via the standard deviations σ_{fd}

$$h = \sqrt{\sum_d \frac{(f_{1d} - f_{2d})^2}{\sigma_{fd}}} \tag{3.2}$$

In the case of a pairwise data clustering of two image objects to a larger image object, the mean spectral homogeneity of all the image objects decreases. The aim of an optimization procedure must therefore be to minimize the reduction of the homogeneity associated with each clustering. An image object should be clustered with the neighbouring segment with which there is the lowest reduction of homogeneity.

We define the *degree of fitting* of two adjacent image objects (h_1 and h_2) via the change of homogeneity h_{diff} before and after a virtual merger (h_m). In accordance with the above-mentioned aim of the optimization procedure, the reduction of homogeneity (h_{diff}) after fusion of the segments should be minimal. This change in homogeneity (h_{diff}) can be established by determining the difference.

---

[1] The distance between two objects can be quantified by the difference. In the following discussion the two terms are used synonymously.

$$h_{diff} = h_m - \frac{h_1 + h_2}{2} \tag{3.3}$$

An improvement can be achieved by adding the image object sizes (n), represented by the sum of the pixels of image object 1 (n_1) and image object 2 (n_2).

$$h_{diff} = h_m - \frac{h_1 n_1 + h_2 n_2}{n_1 + n_2} \tag{3.4}$$

An alternative, also applied in the case of the eCognition software used here, is to weight the homogeneity by the size of the image objects.

$$h_{diff} = h_m(n_1 + n_2) - (h_1 n_1 + h_2 n_2) = n_1(h_m - h_1) + n_2(h_m - h_2) \tag{3.5}$$

For a data set with K channels, where each channel is taken into consideration in accordance with a weighting w_k, this definition can be written as:

$$h_{diff} = \sum_k w_k(n_1(h_{mk} - h_{h1k}) + n_2(h_{mk} - h_{2k})), \quad \text{mit} \quad k = 1, \ldots, K. \tag{3.6}$$

Equations (3.3)–(3.6) therefore evaluate the change in homogeneity h during a fusion of image objects. If this change exceeds a certain threshold value (*last degree of fitting*) then the fusion is not performed. In contrast, if the *degree of fitting* is below the last degree of fitting then the image objects are clustered to form a larger segment. The segmentation procedure stops when no further fusions are possible without exceeding the threshold value. In eCognition, this threshold value is laid down by the user at the start of each segmentation. It is termed the *scale parameter* since a higher threshold value permits more fusions and thus the size of the image objects/segments also increases.

The segmentation procedure uses the following definitions for object homogeneity:

- the color homogeneity (*color*): sum of the standard deviations of the spectral values of the object pixels over all the channels
- the shape homogeneity (*shape*): relative deviation of the edge length from the most compact shape (circle). This is defined in two ways:

  1. Edge criterion (*smoothness*): ratio between the object edge length (l) and the edge length of a bounding box (b)[2]:

$$h = \frac{l}{b} \tag{3.7}$$

[2] A bounding box is the smallest rectangular box around an object that it completely encloses.

2. Area criterion (*compactness*): The deviation from the ideal shape, given by the ratio between the object length (l) and the root of the object size (n) in pixels:

$$h = \frac{l}{\sqrt{n}} \tag{3.8}$$

The shape homogeneity h_f (*shape*) is composed of the two criteria of *smoothness* (S) and *compactness* (C) as follows:

$$h_f = (1 - w_f)S + w_f C, \quad \text{mit} \quad w_f = \text{Gewicht}\,(h_f) \tag{3.9}$$

The homogeneity h for each image object results from the color homogeneity h_c and the shape homogeneity h_f.

$$h = (1 - w_c)h_c + w_c h_f, \quad \text{mit} \quad w_c = \text{Gewicht}\,(h_c) \tag{3.10}$$

These criteria can be interactively selected by the user in eCognition before each segmentation (cf. Fig. 3.2). Furthermore, the required proximity (4th neighbourhood [standard setting] or 8th neighbourhood [taking the corners into consideration]) can be adjusted as well as the scale parameters. As already discussed above, this determines the last degree of fitting and thus indirectly the size of the objects.

3.2.2 Optimization Procedure

In accordance with the definition of homogeneity given in the homogeneity criteria, those objects are to be fused that go together best (*best fitting*). Various decision heuristics are possible in principle, which on the basis of the homogeneity criteria give the degree of fitting with respect to the fusion of two segments.

Starting from any two neighbouring objects, A and B, the following decision heuristics can be distinguished:

1. *Fitting*: Fuse A with any neighbouring object B that fulfils the homogeneity criteria;
2. *Best Fitting*: Fuse A with the neighbouring object B that best fulfils the homogeneity criteria, i.e. that minimizes the change of homogeneity (cf. formulae (3.1)–(3.6);
3. *Local Mutual Best Fitting*: Find for A the neighbouring object B that best fulfils the homogeneity criteria (*best fitting*). Now find for B the neighbouring object C, for which B best fulfils the homogeneity criteria. If $C = A$ fuse the objects, otherwise repeat the procedure with respect to B for A and C for B;
4. *Global Mutual Best Fitting*: Fuse the pair of neighbouring objects that best fulfils the homogeneity criteria in the entire scene.

Simultaneous, or approximately simultaneous, growth is necessary to achieve uniform growth of the segments. This can be achieved by the following conditions:

1. Treat each pixel or object once per cycle;
2. Distribute the completed fusions as far apart as possible over the entire image.

The cycle of a multiscale segmentation makes use of a so-called *distributed sequence*. This sequence systematically processes those points that are at the greatest distance from all previously treated points, which is achieved by using a dither matrix (cf. Abmayer 1994). The distributed sequence also increases the reproducibility of the segmentation (cf. Section 2.3) in contrast to a procedure that treats each object in a random sequence.

In multiscale segmentation, the decision heuristic of *local mutual best fitting* is applied. Uniform growth of the segments is achieved by using a distributed sequence. With respect to the establishment of a hierarchical network of segmentation levels, other rules have to be followed in eCognition:

- The object edges follow the edges of the level immediately below;
- The segmentation is governed by the object edges of the level immediately above.

As already shown, multiscale segmentation has decisive prerequisites and advantages for remote sensing and also for work with high-resolution satellite images.

Problems arise in the attempt to segment image data from different times of imaging in a standardized manner so that the image objects can be compared with respect to actual changes in real-world objects since different object boundaries of the image objects resulting from segmentation can lead to a misclassification. Apart from this problem of independent segmentation, which has already been discussed, there are also freely selectable segmentation parameters which ensure great flexibility. This positive flexibilization of the algorithm can be contrasted with a potential arbitrariness. It is often very time-consuming to select the parameters in such a way that the real-world objects are well demarcated in the subsequent segmentation. The user has no choice but to discover the best parameters by trial and error.

Furthermore, it would be desirable to have, for example, predefined settings in order to selectively extract objects in certain spatial data. Recent work by Feitosa *et al.* (2006) is moving in this direction.

3.3 Semantic Modelling

The semantic modelling in eCognition consists of a feature analysis for the image objects obtained in the multiscale segmentation (cf. Section 3.2) with the subsequent formation of a network of rule-based semantic features (cf. Section 2.3). In eCognition there are a large number of predefined features for describing object properties. Furthermore, user-defined features, so-called customized features can be set up. For an overview of these predefined features see Baatz *et al.* (2004).

The membership of image objects in object classes is defined via so-called *membership functions* of the features (cf. Fig. 3.5). This means that depending on the feature intensity of an image object it is a member of a certain object class. The dynamic range of these functions is [0, 1] in eCognition and is plotted on the *y*-axis. The *x*-axis indicates the respective feature intensity. Membership can either be defined as a fixed threshold value or as a fuzzy logic threshold value. In the case of fixed threshold values, an image object either belongs to object class [1] or it does not [0]. The *membership functions* functions 3, 4, 9 and 12 from Fig. 3.5 can be used here. In the case of a fuzzy logic threshold value, membership of a class varies depending on the feature intensity so that an object may belong to two object classes. The object is then assigned to that object class to which it has the highest membership value. Typical membership functions based on fuzzy logic are the shapes 1, 2, 5 and 6 in Fig. 3.5. The fuzzy logic theory is founded on the fuzzy set theory derived by Lofti Zadeh in 1965 (cf. Section 3.4). Twelve membership functions are already predefined in eCognition 4.0. Furthermore, any functions required can be interactively set up and modified.

The feature analysis has the task of identifying characteristic features for the individual object classes. Once these typical features have been determined classification rules are drawn up for each object class. In the present study, feature analysis is performed on the basis of the developed SEaTH methodology (cf. Chapter 5). The object classes defined via classification rules by means of

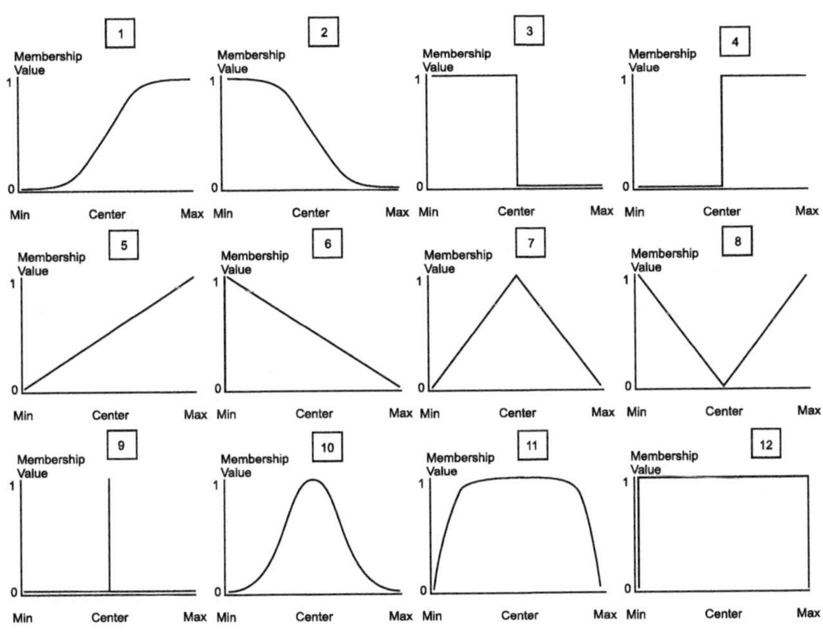

Fig. 3.5 Membership functions in eCognition 4.0

feature intensity are combined to form a hierarchical network, the class hierarchy. The class hierarchy has the following possible relations.

- (*Inheritance*): permits the formation of *parent classes* and *child classes*. The parent class passes on its feature descriptions to the children (1st generation), who can themselves also have child classes (2nd generation). The 2nd generation inherits the features of the parents and the first generation of children.
- *Groups*: classes which have a semantic relation can be combined to form groups.
- *Structure*: Object classes can be combined to form structural groups as a basis for the classification-based segmentation.

A great advantage is the large number of predefined features in eCognition. A good description of the object classes by characteristic features should thus be possible. Unfortunately the software does not have a comprehensive tool which can analyse this large number of possible features fast, effectively and also statistically. Limited possibilities of identifying features for classification are available but they are not sufficient for effective, automated semantic modelling. An analysis is very tedious and time-consuming using these tools alone and the user is never certain whether this is the optimum model with optimum threshold values. The lack of a statistical feature extraction and analysis tool is one of the weak points of eCognition since the potential of a broad feature basis cannot be fully exploited.

3.4 Semantic Classification

After multiscale segmentation with subsequent semantic modelling, the semantic classification is performed, dividing the image objects into object classes. The objects can be assigned to classes by a nearest *neighbour classifier* or by a *classification model based on classification rules*.

With the nearest neighbour classifier, an unknown object is assigned to the object class to whose centre-of-gravity vector in the feature space it is closest. First of all, training areas are assigned manually to each object class. The centre-of-gravity vectors of the individual object classes are obtained by taking the mean of the feature values of these training areas. In this classification, all objects present are classified. Since this classification is a standard technique, more detailed information can be found, for example, in Kraus (1990), Mather (2004) or Schowengerdt (1997).

In the present study, use was exclusively made of the rule-based classification model. In semantic modelling, the representative features of the object classes are determined with the associated threshold value and implemented in a semantic model. This is then the basis for the classification. Membership of objects in certain classes is then regulated via classification rules with fixed threshold values. In general, fixed threshold values can determine class membership or else threshold values based on fuzzy logic. A fuzzy set A is characterized by its membership function μ_A, which assigns to every element x in the reference set X a real number in a closed interval (0, 1) (Tizhoosh, 1998).

$$A = (x, \mu_A(x))|x \in X \tag{3.11}$$

This contrasts with the classical logical sets in which only elements with the membership value of $0 or 1$ can be given, i.e. whether the object belongs to the class described (value = 1) or not (value = 0). In fuzzy set theory, on the other hand, the object may be a member of various classes. It is finally assigned to the object class with which it has the greatest membership. For more detailed information on fuzzy set theory see Nauck *et al.* (1994) and Zadeh (1965).

In this study, a classification model with fixed threshold values is used for the classification. If, for example, an object exceeds a certain boundary in the feature under consideration then it no longer belongs to the described object class. The modelled features and threshold values are statistically optimal due to the analysis with SEaTH (cf. Chapter 5) for feature extraction. With the later object-based change detection fuzzy logic classification rules can be used to determine those change components that have the greatest intensity in the object under consideration (cf. Section 8.1.2).

A great advantage of a rule-based classification model is the transferability option. In this way, a reference model can be set up which can later be transferred rapidly and simply to any satellite scene. The quality of the results obtained and what differences there are in spatial and temporal transferability will be investigated in Chapter 7.

Chapter 4
Change Detection

Change detection is generally considered to mean the identification of changes in image scenes taken at different points in time. Changes in the earth's surface can be identified by remote sensing. Due to the nature of the existing satellite scanners, in past decades large-scale features, such as forest cover or land use, have mainly been studied. With increasing resolution of the satellite sensors, smaller-scale phenomena can now be discovered and investigated. There are currently still no specific methods for the field of object-based change detection. However, first attempts have been made, as in work by Blaschke (2005), to detect changes going beyond object boundaries. Since it is as yet not possible to independently reproduce the segmentation results, it still remains to be seen whether the approach proves promising beyond object boundaries. In the present study, the new method of combined change detection will be presented in Chapter 8. This combines a pixel-based change detection technique with an object-based image analysis. In the following, some basic techniques of pixel-based change detection will be briefly outlined and then the method of "multivariate alteration detection" used in this study is derived and described in detail.

The change detection in the present study is performed against the background of nuclear safeguards and nuclear verification (cf. Section 1.1). Changes in artificial objects in and around nuclear facilities such as buildings, fences or roads will be detected and interpreted. A measure of change is usually obtained by comparing reflectance values. In this way, the spectral values of each pixel are compared at one point in time with its spectrum at a different point in time. A measure of change can then be derived from the similarity or discrepancy of the spectra.

Within the framework of applications for nuclear verification purposes, change detection is a suitable tool for monitoring compliance with commitments and treaties. With respect to nuclear safeguards, it can be used to identify undeclared structural changes in facilities or to detect newly constructed plants or structures such as buildings for the first time.

S. Nussbaum, G. Menz, *Object-Based Image Analysis and Treaty Verification*,

© Springer Science+Business Media B.V. 2008

4.1 Basic Techniques

4.1.1 Arithmetic Operations

The relatively simple techniques of change detection include arithmetical operations such as difference and quota participation or the difference of indices (Jensen, 2004; Mas, 1999; Singh, 1989).

If the changes in two multispectral images represented by the N-dimensional vectors **F** and **G** are sought, then they can be found by simply determining the *difference*. The individual vectors are subtracted channel by channel.

$$\mathbf{F} - \mathbf{G} = (F_1 - G_1, F_2 - G_2, \ldots, F_N - G_N)^T \tag{4.1}$$

Pixel values close to zero in the difference image indicate that there has been no change, whereas large positive or negative values point to a possible change. Significant changes can be defined via the choice of threshold values, usually expressed by the standard deviations σ. As an alternative to differences, *quotients* of the pixel intensity can also be formed:

$$\frac{F_K}{G_K} \qquad with \qquad k = 1 \ldots N \tag{4.2}$$

Since these procedures involve simple and readily applicable methods they have been used in a number of papers such as Fung (1990), Macleod and Congalton (1998), Muchoney and Haack (1994), Ridd and Liu (1998), Sohl (1999) or Sunar (1998).

Arithmetic operations are not only conceivable with individual pixel values. The *differences or quotients of indices* can also be determined. Examples to be mentioned here are the well-known NDVI (*Normalized Differenced Vegetation Index*) and the NDASI (*Normalized Differenced Anthropogenic Surface Index*). These indices can, however, only be used to track down certain types of change and only for certain types of satellite sensor.

Changes in a satellite scene are expressed in an altered reflectance value of the pixels. In a bi-dimensional feature space, the *change vector analysis* spans a vector between the pixel value of the first point in time t_1 and the pixel value of the second point in time t_1. The changes can be described by the direction and length of the vector and can be divided into classes of similar changes.

4.1.2 Comparison of Classifications

Another method of change detection is a comparison according to a classification of the images under consideration. In this way, all the monitored and unmonitored classification processes normally used in remote sensing can theoretically be applied. Possibilities of interpreting the classification once again arise from arithmetic operations, such as determining the difference. Another alternative is the *change detection matrix* (Sohl, 1999), by means of which transitions from one class to the next can be illustrated.

The comparison of classification results is typically made on a pixel basis. Errors in the classification have a negative effect on the result of change detection and are superimposed on changes that may be real. In the object-based analysis, as already mentioned, imprecise segmentations also lead to an incorrect classification of changes.

4.1.3 (Iterative) Principal Component Transformation

Principal component analysis (PCA) is a method frequently used for concentrating information in remote sensing. This method can also be applied for the field of change detection, as will be briefly described in the following. The PCA will not be derived and discussed in detail instead attention is drawn to the standard literature such as Mather (2004), Richards and Jia (1999) or Schowengerdt (1997). This method has already been derived and discussed in detail in these and numerous other works.

For bitemporal multispectral images, a two-dimensional feature space (X) can be generated for each spectral channel (K). In this feature space the position of each pixel is given by a point (f_K^{t1}, f_K^{t2}), relative to the random vector ($\mathbf{F}_K^{t1}$, $\mathbf{F}_K^{t2}$). Since the *no-change* pixels from both points in time are closely correlated they cluster along the first principal component axis (PC1) whereas the *change* pixels are positioned away from this axis in the feature space. A spectral change is revealed by a deviation vertical to PC1, that is to say by the second main component (PC2). PC2 consequently permits a quantification of the degree of change for a given pixel (cf. Fig. 4.1).

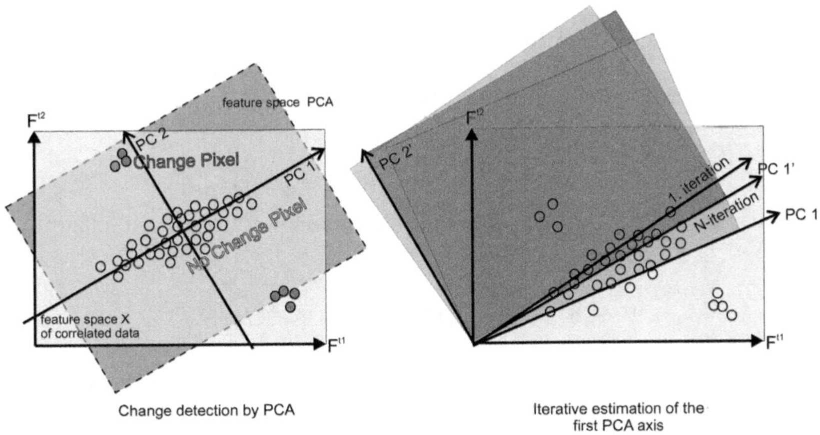

Fig. 4.1 Change detection with the (iterative) principal component transformation

PCA is still a frequently used technique for change detection (Byrne *et al.*, 1980). However, there have been variations in the methodology. Thus, for example, PCA can be performed with nonstandardized or standardized variables (Eklundh and Singh, 1993; Singh, 1989; Fung and LeDrew, 1987). Furthermore, it is possible to apply PCA to difference images (Gong, 1993), to use indices (Ingebritsen and Lyon, 1985) or to perform PCA on a channel-by-channel basis (Richards and Jia, 1999; Jiaju, 1988; Chavez and MacKinnon, 1994; Mas, 1999). Another more recent method is the estimation of the main axes without a covariance or correlation matrix. In this case, neural networks are used (Canty, 1999; Xu and Yuille, 1995; Yang and Wang, 1999).

A *Principal component analysis* determination of the first principal axis (PC1) through all the pixels (change and no change) may prove to be problematic. In order to avoid this, it is possible to make an iterative estimate of the principal components (Wiemker, 1997; Wiemker *et al.*, 1997). All the pixels continue to be considered in each iteration, but weighted by their current probability of being a no-change pixel. The probability is thus given by the distance of the pixels from the presently estimated position of the first principal component (PC1). After a few iterations the result finally converges around PC1 (cf. Fig. 4.1).

4.2 Multivariate Alteration Detection – MAD

A more recent method for change detection with multispectral, bitemporal satellite images is multivariate alteration detection (MAD). This is a pixel-based method of linear transformation. The aim of the MAD procedure is to transform two multispectral data sets in such a way that the maximum change (variance) of all channels can be seen *at the same time* in a *single* difference image. The MAD procedure mainly derives from the work of Conradsen and Nielsen (Nielsen, 1996; Nielsen and Conradsen, 1997).

Let a satellite image F be given at a point in time t_1, represented over all channels N by the vector $\mathbf{F}$. A single image (U) can be generated from this vector via a linear transformation having the following pixel intensities:

$$U = \mathbf{a}^T * \mathbf{F} = a_1 F_1 + a_2 F_2 + \ldots a_N F_N. \tag{4.3}$$

The same is true of the image vector $\mathbf{G}$ for the point in time t_2. As a linear transformation (V):

$$V = \mathbf{b}^T * \mathbf{G} = b_1 G_1 + b_2 G_2 + \ldots b_N G_N. \tag{4.4}$$

Finally, the change image is obtained from the difference of the images $U - V$.

The choice of the transformation coefficient vectors $\mathbf{a}^T$ and $\mathbf{b}^T$ has a decisive influence on the result of this change detection. At the moment these vectors have not yet been determined. Nielsen *et al.* (1998) propose that the coefficients should be selected in such a way that the positive correlation between U and V is minimized. In other words, the change image $(U - V)$ should have a maximal variance. Assuming that the variance of the two scenes mainly results from current changes in the period $t_2 - t_1$, these changes are emphasized as far as possible by the MAD procedure.

Those linear combinations are therefore sought that maximizes the variance:

$$var(U - V) = var(U) + var(V) - 2cov(U, V) \longrightarrow Maximum, \tag{4.5}$$

on condition that

$$var(U) = var(V) = 1. \tag{4.6}$$

For the first condition this leads to

$$var(U - V) = 2(1 - \rho), \tag{4.7}$$

where ρ is the correlation of the transformed vectors U and V,

$$\rho = corr(U, V) = \frac{cov(U, V)}{\sqrt{var(U)var()V}}. \tag{4.8}$$

Since changes between two periods of time are intended to be highlighted, it is furthermore required that the random variables U and V should be positively correlated with each other so that the following holds:

$$cov(U, V) > 0. \tag{4.9}$$

According to the proposal by Nielsen *et al.*, the vectors **a** and **b** are now determined, which minimize the positive correlation ρ.

The transformation vectors are determined by a canonical correlation analysis, whose mathematical derivation will not be shown here. This step is described in detail in Nielsen (1996), Nielsen and Conradsen (1997). It was found that vectors **a** and **b** can be determined via the *generalized eigenvalue problem*.

$$\Sigma_{fg}\Sigma_{gg}^{-1}\Sigma_{gf}\mathbf{a} = \rho^2\Sigma_{ff}\mathbf{a} \tag{4.10}$$

$$\Sigma_{gf}\Sigma_{ff}^{-1}\Sigma_{fg}\mathbf{b} = \rho^2\Sigma_{gg}\mathbf{b} \tag{4.11}$$

This thus gives the desired vectors of the transformation $U = \mathbf{a}^T\mathbf{F}$ via the eigenvectors $\mathbf{a}_1^T \ldots \mathbf{a}_N^T$, in accordance with the general eigenvalues

$$\rho^2\lambda^1 \geq \ldots \geq \lambda^N \tag{4.12}$$

of $\Sigma_{fg}\Sigma_{gg}^{-1}\Sigma_{gf}$ with respect to Σ_{ff}. The same is true of the transformation $V = \mathbf{b}^T\mathbf{G}$ where the desired vectors are given by the eigenvectors $\mathbf{b}_1^T \ldots \mathbf{b}_N^T$ of $\Sigma_{gf}\Sigma_{ff}^{-1}\Sigma_{fg}$ with respect to Σ_{gg}. Consequently, N difference images result. Nielsen *et al.* call these difference components the *MAD components* of the bitemporal initial data.

$$MAD_i = U_i - V_i = \mathbf{a}_i^T\mathbf{F} - \mathbf{b}_i^T\mathbf{G}, \qquad i = 1 \ldots N \tag{4.13}$$

The transformation belonging to the smallest eigenvalue (a_N^T, b_N^T) contains the greatest variance for the difference $U - V$, since the MAD components in the feature space are arranged orthogonally to each other and thus are uncorrelated.

In order to prevent any change however small, for example a slight change in reflectance from a roof due to a different time of day, leading to the image being classified as a change there is a need to assign threshold values. The threshold values then determine whether a pixel is regarded as a significant *change* or not (*no change*). Since the MAD components have a Gaussian distribution and are greater than zero and also are not correlated with respect to each other, the threshold value can be determined via the standard deviation σ. The threshold value is determined separately for a predefined deviation from the mean for each MAD component. The advantage of using standard deviations for the threshold values is that it provides information on the significance level of the changes. A threshold value of 2σ, for example, means a change on the 95% confidence level .

The threshold value can also be automatically determined on the basis of the image content. Fung and LeDrew (1988) use an iterative method according to which the threshold is increased or reduced by 0.1 σ after each iteration. In order to determine the level of detail they calculate the overall accuracy and the kappa index. Gong (1993) and Metternicht (1999) propose a fuzzy membership function in order to create a threshold value in change detection.

The approach of Bruzzone and Prieto (1999) used in the present study for the SEaTH methodology (cf. Chapter 5) can also be employed for determining the threshold value. It is, however, assumed that the changes also have a Gaussian distribution. The approach uses two Bayesian strategies for the automatic creation of threshold values for difference images. The methodology is based on the assumption that a probability density function $p(X)$ of the grey values of the difference image D can be modelled as a combined density of the classes *no change (NC)*, *change+* (C+) and *change- (C−)* with a Gaussian distribution.

$$p(X) = p(X|NC)p(NC) + p(x|C-)p(C-) + p(x|C+)p(C+) \tag{4.14}$$

The procedure, which consists of the three stages of initialization, iterative estimate and creation of threshold values, makes use of the following Bayesian strategies: the Bayes rule for minimum error (BRME) and the Bayes rule for minimum cost (BRMC). The mathematical derivation and also more detailed information can be found in the articles specified above.

A great advantage of the MAD transformation is that it is invariant with respect to linear changes. This also involves a certain insensitivity to atmospheric differences, sensor inaccuracies and radiometric deviations in the RS data investigated since these types of changes often proceed linearly. Changes in the external parameters of a image, such as those described above, often conceal real changes in the earth's surface. A method that is insensitive to this type of change is of great benefit, especially for the field of high-resolution satellite images and in the case of great temporal distance of the images considered.

The MAD components indicate the significant changes over time. On the other hand, it is not possible to determine what has changed by using the MAD method alone.

In her work, Niemeyer (2001) comes to the conclusion that the MAD procedure is on the whole very well suited for establishing significant changes. The conclusion was drawn against the background of nuclear verification with satellite images (moderate resolution). Since nuclear verification is also the subject of the present study, the MAD technique is regarded as a promising method for change detection.

The present study will investigate the extent to which a combination of pixel-based MAD with object-based image analysis can improve possibilities for interpreting the changes detected. It is, for example, conceivable that object-based classification and MAD change information could be combined. These two levels can be linked via the eCognition software. In this way, the type of change can be precisely determined. These and other considerations and approaches will be discussed in Chapter 8.

Part II
Object-Based Analysis and Change Detection

Chapter 5
SEaTH – A New Tool for Feature Analysis

A decisive and complex point in object-based, rule-based image analysis is feature analysis, which will be discussed in detail in the following.

After the initial data sets have been segmented, a data set consisting of image objects is available for further analysis. The number of image objects is different in each segmentation level created by eCognition (see Section 3.2 and also Fig. 3.4). In this way, a low level has more image objects than a higher level. The present study uses data sets in the order of 3000×3000 pixels. In order to be able to satisfactorily define all real-world objects as image segments several segmentation levels are required with between approx. 400 and 40,000 image objects. For each individual image object, eCognition provides more than 150 object properties, which are also known as object features. This includes spectral, shape, texture and also proximity features.

The attempt to simply use all existing features for classification purposes, for example via a simple nearest neighbor classifier, leads to a very poor result due to the hundredfold dimensionality of the feature space. Consequently, features must be selected for classification purposes. Each object class has characteristic features enabling it to be separated from other object classes to a greater or lesser extent.

This brief description of the problem shows that a good feature analysis should be able to analyse a large number of features, it should identify the characteristic features for any number of object classes and it should, moreover, determine the threshold for the feature intensity which achieves optimum separation from the other object classes. In order to be able the find the best feature for the classification and to make individual features distinguishable from each other, a uniform measure of quality for features should be defined.

A good feature analysis is a basic prerequisite for successful work in object-based image analysis. At present, such a comprehensive feature analysis has not yet been implemented in the eCognition software. In most cases, feature analysis is performed empirically by trial and error. In this way, certain features and thresholds are tested, some of which are already known, and are confirmed or rejected on the basis of the classification result obtained. For reasons of time and effort is impossible to examine all possible features so that the selected threshold is not statistically optimized and no conclusion can be drawn as to the quality of the features. Other

S. Nussbaum, G. Menz, *Object-Based Image Analysis and Treaty Verification*,

© Springer Science+Business Media B.V. 2008

features might have led to a better result so that the empirical approach frequently does not lead to an optimal result.

Another problem arises for verification in connection with the NPT. As already mentioned, a priori knowledge on the part of the user can be of significance in the feature analysis. Jasani (2004), for instance, investigated German power plant complexes and general indications of features can be obtained from this study. In the present study, only a priori knowledge of the Bushehr power plant can be integrated into the analysis. The attempt will, however, be made to transfer the key features identified by Jasani to this plant. No reliable a priori knowledge is available for any of the other study areas, and in particular not for the Esfahan nuclear site.

With respect to further work, this therefore means that a reliable result must be achieved with little additional information. Available information is restricted to data from remote sensing, since on-site inspections in the study area are impossible at the moment for political reasons. The characteristic features of the object classes sought must therefore be extracted from the satellite data themselves.

A new method known as SEaTH has been developed in the present study. This tool is the first approach towards a comprehensive feature analysis. SEaTH fulfils all the above-mentioned requirements for good feature analysis. With SEaTH it is possible not only to extract characteristic features but also the associated thresholds for any number of object classes from the large number of object features available, and the result of the SEaTH analysis, the separability, is at the same time a comparative measure of quality for the features studied.

5.1 SEaTH Method

The feature analyzing tool SEaTH (*SEparability and THresholds*) identifies these characteristic features with a statistical approach based on training objects. These training objects represent a small subset out of the total amount of image objects and should be representative objects for each object class. The statistical measure for determining the representative features for each object class is the pairwise separability of the object classes among each other. Subsequently, SEaTH calculates the thresholds which allow the *maximum* separability in the chosen features.

The identification of the characteristic features is a problem of probability density estimation. On the basis of representative training data for each object class, the probability distribution for each class can be estimated and used to calculate the separability between two object classes. Under the assumption of normal probability distributions, the *Bhattacharyya distance* B can be used as a suitable separability measure. B is justified as a measure of separability from the Bayesian decision rule for misclassification probability. For the derivation of the Bhattacharyya distance see Bhattacharyya (1943) or Fukunaga (1990). For two classes (C_1, C_2) and one feature it is given by

$$B = \frac{1}{8}(m_1 - m_2)^2 \frac{2}{\sigma_1^2 + \sigma_2^2} + \frac{1}{2} ln[\frac{\sigma_1^2 + \sigma_2^2}{2\sigma_1\sigma_2}], \tag{5.1}$$

where m_i and σ_i^2, $i = 1, 2$, are the mean and the variance, respectively, for the two feature distributions. If the means coincide, the first term in (5.1) vanishes, whereas the second term vanishes if the two feature distributions have equal variances.

Figure 5.1 shows the probability distribution exemplified for two object classes C_1 and C_2 and three notional features A, B and C. In feature A both object classes show a *partial separability*, this means that there is an area where the probability distributions of the object classes (C_1 and C_2) overlap in their feature characteristic. Given feature B this overlap is so large that its use for classification would result in a huge object misclassification rate. This feature therefore provides *poor separability* relative to object classes C_1 and C_2. The ideal case is represented by feature C. Here the object classes have no overlap in the feature characteristic it is therefore well-suited for classification: the feature has *complete separability*.

An more useful measure for separation in classification contexts is the *Jeffries–Matusita distance* J which has, unlike B, a finite dynamic range. This allows a better comparison of the feature analysis results to identify that feature which has the best separability. The Jeffries–Matusita (Richards and Jia, 1999) distance measures the separability of two classes on a scale $[0 - 2]$ in terms of B:

$$J = 2(1 - e^{-B}). \tag{5.2}$$

Complete separability of the two classes with respect to the analyzed feature is indicated by $J = 2$. That is to say, on the basis of the training objects used, there will be no misclassifications if this feature is used for classification. The lower J is, the worse is the separability and the higher the number of misclassified objects. SEaTH calculates the separability for any number of given object classes and object class combinations.

Besides determining the features separating optimally the object classes among each other, it is essential to know also the decision threshold for the maximum separability. The knowledge of the optimum threshold is necessary for the assembly of a ruled-based classification model.

The optimum threshold is also calculated by SEaTH. A Gaussian probability mixture model of the form

$$p(x) = p(x|C1)p(C1) + p(x|C2)p(C2)$$

is fit to the frequency distribution of a feature for two object classes $C1$ and $C2$, where $p(x|C1)$ is a normal distribution with mean m_{C1} and variance σ_{C1}^2 and similarly for $p(x|C2)$. The decision threshold which minimizes the error probability is obtained by solving

$$p(x|C1)p(C1) = p(x|C2)p(C2). \tag{5.3}$$

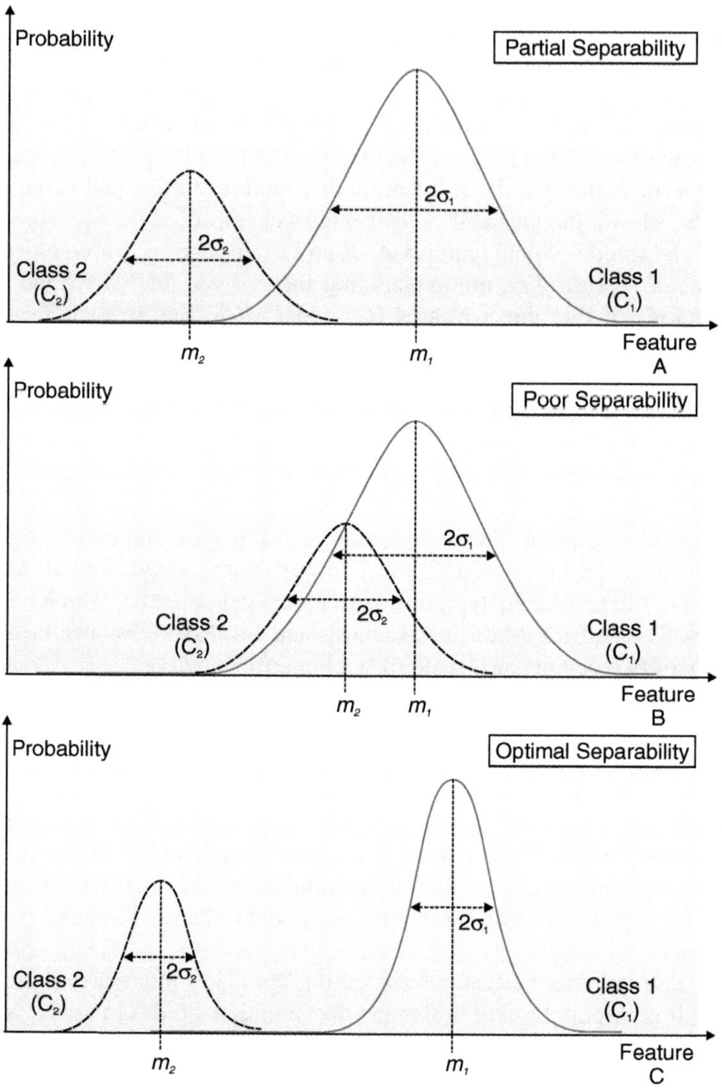

Fig. 5.1 Examples of probability distributions

for x. Taking logarithms,

$$\frac{1}{2\sigma_{C2}^2}(x - m_{C2})^2 - \frac{1}{2\sigma_{C1}^2}(x - m_{C1})^2 = log[\frac{\sigma_{C1}}{\sigma_{C2}} * \frac{p(C2)}{p(C1)}] =: A \tag{5.4}$$

with solutions

$$x_{1(2)} = \frac{1}{\sigma_{C1}^2 - \sigma_{C2}^2}\left[m_{C2}\sigma_{C1}^2 - m_{C1}\sigma_{C2}^2 \pm \sigma_{C1}\sqrt{(m_{C1} - m_{C2})^2 + 2A(\sigma_{C1}^2 - \sigma_{C2}^2)} \right]. \tag{5.5}$$

The relevant solution of the two can be determined by requiring that it lie between the two means m_1, m_2 of the probability distributions. Thus, for the example in Fig. 5.2, x_1 is the correct choice. Since the distributions are only *partially separated*, there will be some misclassifications when using this feature for classification of unknown object classes. Given the validity of the normal approximation assumption, SEaTH will minimize their number. If the probabilities are not normally distributed, the calculated separability for this feature is low, i.e. it will not be used for classification purposes. This thus ensures that only very good features are taken into consideration. Work is currently in progress for features which have probabilities that are not normally distributed since these features are, of course, also potentially suitable for classification purposes.

To identify the best features for the classification SEaTH calculates the separability and the corresponding threshold for every object class combination and for every feature. Any number of object classes and features can be analyzed. The results of SEaTH are presented in tables. An interpretation of the results allows a fast preparation of a classification model, with statistically optimized features and thresholds. SEaTH makes it possible to analyse a large number of features for object description in a short time. From Chapter 7 onwards this method is applied for feature extraction. In the following, the SEaTH methodology is validated using the example of an aerial photograph of Research Centre Jülich.

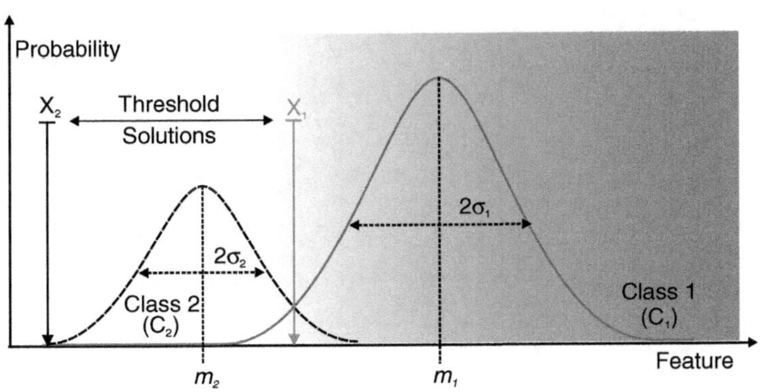

Fig. 5.2 Threshold identification

5.2 SEaTH Validation

Validation of the SEaTH methodology is performed by evaluating the quality of an object-based image analysis. The classification is made on the basis of an aerial photograph of Research Centre Jülich (FZJ) taken in September 1998 with a spatial resolution of 1.25 m (copyright Geobasisdaten: Landesvermessungsamt NRW). The Research Centre was selected as a model area since it contains ground information on all the reference objects selected. All image objects present are used as reference objects for evaluating the quality, with the exception of those image objects that include mixed use due to the segmentation within their segment boundaries. Figure 5.3 shows the aerial photograph of Research Centre Jülich.

The image analysis process was carried out with the software eCognition. The object classes of **buildings**, **streets**, **meadows**, **forest**, **fields** and **shadow** will be identified in the scene. The class buildings includes buildings as well as sealed areas that are not streets. Object extraction is performed by segmenting the image. A good object extraction for all selected object classes is achieved with a *scale factor* of 50. The *shape factor* is weighted with 0.3, and is subdivided into *compactness* (0.5) and *smoothness* (0.5). This most general form of segmentation was selected for validating the methodology. This general, non-hierarchical type of segmentation results in objects that contain mixed use and therefore cannot be clearly assigned to one object class. However, these objects are in a minority and are not taken into consideration in selecting training objects nor are they used as reference objects. It

Fig. 5.3 Aerial photograph of Research Centre Jülich (FZJ)

may be assumed that an improved segmentation would only improve the validation results, i.e. the classification result, and not make them worse.

Training objects are defined in eCognition and analysed with SEaTH for the object classes. Spectral, shape and textural features are included in the feature analysis. Of the more than 3000 image objects available, about 3% were selected as training objects.

SEaTH analysis on the basis of 71 selected features (see Table 5.1) every possible combination of object classes and lists the results of the calculations in the form of tables. For reasons of space, Table 5.2 shows a summary of the result. Only the two "best" features in the separability criterion per class combination are shown for selected combinations of classes. As a rule, it is not necessary to use many more than two features for classification purposes. In view of a later possible transferability of the classification model, it also makes sense to use a small number of characteristic features for modelling. An interpretation of the SEaTH result leads to the compilation of a rule-based classification model.

Table 5.1 Features used for image analysis with SEaTH

Spectral features
Mean *Channel 1,2,3*
Stddev *Channel 1,2,3*
Ratio *Channel 1,2,3*
Max.Diff.
Brightness
Shape features
Area (m^2)
Length (m)
Width (m)
Length/width
Compactness
Elliptic Fit
Rectangular Fit
Border length (m)
Shape index
Density
Main direction
Asymmetry
Texture features
GLCM Homogeneity (all dir.) *Channel 1,2,3, all dir.*
GLCM Contrast (all dir.) *Channel 1,2,3, all dir.*
GLCM Dissimilarity (all dir.) *Channel 1,2,3, all dir.*
GLCM Entropy (all dir.) *Channel 1,2,3, all dir.*
GLCM Ang. 2nd moment (all dir.) *Channel 1,2,3, all dir.*
GLCM Mean (all dir.) *Channel 1,2,3, all dir.*
GLCM StdDev (all dir.) *Channel 1,2,3, all dir.*
GLCM Correlation (all dir.) *Channel 1,2,3, all dir.*
GLDV Ang. 2nd moment (all dir.) *Channel 1,2,3, all dir.*
GLDV Entropy (all dir.) *Channel 1,2,3, all dir.*
GLDV Mean (all dir.) *Channel 1,2,3, all dir.*
GLDV Contrast (all dir.) *Channel 1,2,3, all dir.*

Two class combinations will be discussed as an example of how to interpret this table. SEaTH reveals that objects in the **buildings** class can be most easily separated from those in the **forest** class by means of the feature *Ratio Channel 2*. This feature

Table 5.2 Summarized result of the SEaTH analysis for the aerial photograph of FZJ

Object Class Combination	Separability	Omen	Threshold
BUILDINGS and FOREST			
Ratio Chan. 2	1.99	small	0.352
Ratio Chan. 1	1.95	great	0.309
BUILDINGS and STREETS			
Density	1.69	great	1.105
Length (m)	1.17	small	150.901
BUILDINGS and MEADOWS			
Ratio Chan. 2	1.85	small	0.351
Max.Diff.	1.80	small	0.107
BUILDINGS and FIELDS			
Ratio Chan. 3	1.97	great	0.311
Ratio Chan. 1	1.89	small	0.345
SHADOWS from BUILT-UP			
Mean Chan. 2	1.98	small	94.810
Mean Chan. 1	1.98	small	85.900
SHADOWS and FOREST			
GLCM Ang. 2nd (a.d.) Chan. 2	1.41	great	0.0006
GLCM Ang. 2nd moment (all dir.)	1.37	great	0.0007
SHADOWS and STREETS			
Mean Chan. 1	2.00	small	117.769
GLCM Mean (all dir.) Chan. 1	2.00	small	121.413
SHADOWS and MEADOWS			
Mean Chan. 2	1.96	small	93.979
GLCM Mean (all dir.) Chan. 2	1.94	small	99.268
FOREST and STREETS			
GLCM Mean (all dir.) Chan. 1	2.00	small	120.944
Mean Chan. 1	2.00	small	130.044
FOREST and MEADOWS			
GLCM Dissim. (all dir.) Chan. 2	1.99	great	12.920
GLDV Mean (all dir.) Chan. 2	1.99	great	12.920
FOREST and FIELDS			
GLCM Dissim. (all dir.) Chan. 2	1.99	great	13.333
GLDV Mean (all dir.) Chan. 2	1.99	great	13.333
STREETS and FIELDS			
Ratio Chan. 3	1.97	great	0.316
Max.Diff.	1.94	small	0.078
STREETS and MEADOWS			
Ratio Chan. 2	1.92	small	0.344
Max.Diff.	1.89	small	0.074
MEADOWS and FIELDS			
Ratio Chan. 1	1.82	small	0.345
Ratio Chan. 2	1.52	great	0.361

achieves almost complete separability (1.99). The threshold permitted by this "best" separability is 0.352. All image objects with a *Ratio Channel 2* smaller than 0.352 should therefore be assigned to the **buildings** class. This would then be the first classification rule for the classification model. Furthermore, the class combination **forest** and **meadows** indicates that two texture features permit the best separation of these object classes. All image objects with an intensity greater than 12.92 in the feature *GLCM Dissimilarity Chan. 2.* should belong to the object class **forest**. If the intensity is less than this then the objects are **meadows**.

All combinations are to be interpreted in this way and classification rules developed from the interpretation. SEaTH additionally provides quantitative information on the classification to be made and the results to be expected. It can be seen that the class combinations of **buildings** and **forest**, **buildings** and **fields**, **shadow** and **buildings**, **shadow** and **streets**, **shadow** and **meadows**, **forest** and **streets**, **forest** and **meadows**, **forest** and **fields**, and **streets** and **fields** can be almost completely separated since the separability is always more than 1.9. It is more difficult to separate **buildings** from **streets** (1.69) and **shadow** from **forest** (1.41), although both separabilities can still be regarded as good. It is, however, to be expected that in these classes more objects from one class will be incorrectly assigned to the other class. It is also becomes clear that it is not only spectral values that lead to the best separability but also shape and texture features. This speaks in favour of an improved classification result with an object-based classification in contrast to the pixel-based method due to the broader feature basis.

Figure 5.4 shows the classification model developed from the SEaTH analysis. Since there is only one object in its class, the water class is modelled on the basis of a priori knowledge. The object-based classification result based on the model is shown in Fig. 5.5.

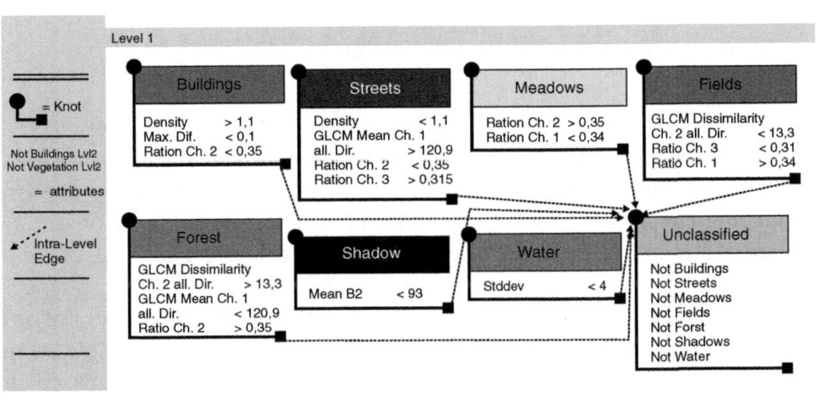

Fig. 5.4 Classification model of FZJ. The individual object classes are shown with the associated classification rules. The dotted arrows represent an internal linkage, which is termed the "intralevel edge" in semantic modelling. The class unclassified is thus simply composed of all image objects that do not belong to any other object class

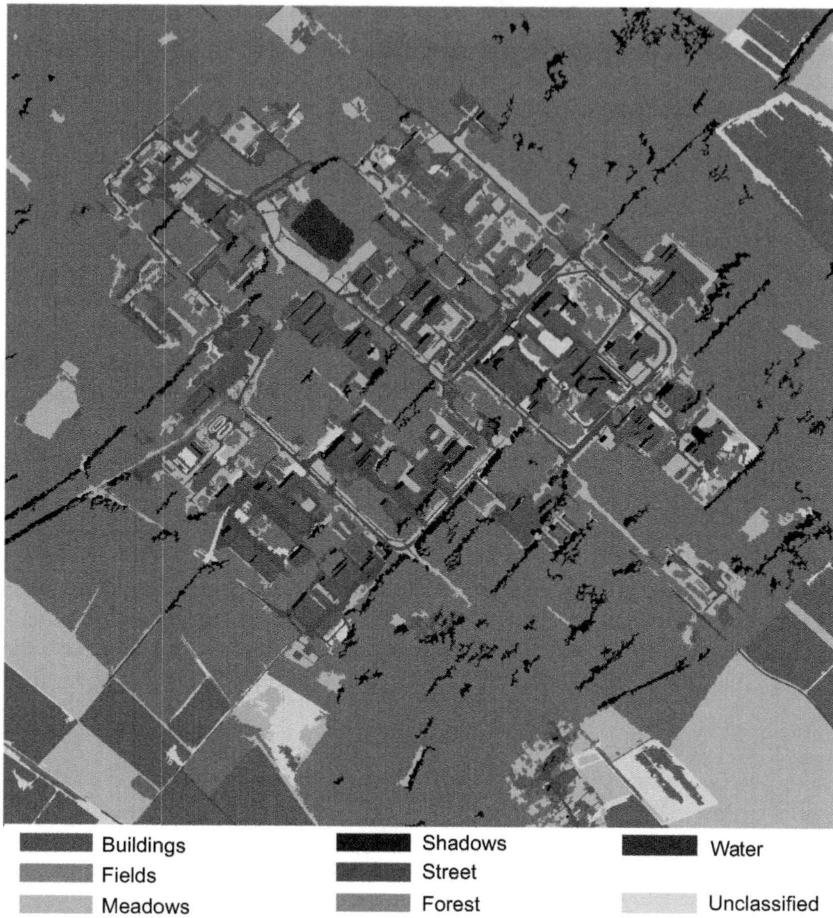

Fig. 5.5 Result of the object-based classification of FZJ

In the following, the quality of the classification is calculated as a validation for the SEaTH method. The precision of spatial data is defined by the *United States Geological Survey (USGS)* as: "The closeness of results of observations, computations, or estimates to the true values or the values accepted as being true" (USGS 1990).

In remote sensing, estimates of precision are an important technique for assessing and comparing classification results. There are various methods of determining this precision which have already been derived and extensively discussed in the literature. Banko (1998), Congalton (1991), Rosenfield *et al.* (1982), Hudson and Ramm (1987) and van Genderen *et al.* (1978) give a good overview of the various measures, the general methodology as well as the establishment of an error matrix and the topic of training areas (samples).

In the present study, the mathematical derivation of the individual measures will not be shown and readers are referred to the literature mentioned above.

For quality assessment (estimates of precision) the *overall accuracy* and also the *user's and producer's accuracy* are calculated. The representation takes the form of a conventional *error matrix*. This matrix lists the *reference data (ground truth)* in the columns and the classified *training areas* in the rows. The diagonal of this matrix shows the correct classified objects.

A *user's and producer's accuracy* basic measure is the *overall accuracy*, which can be calculated from the ratio of the number of correctly classified samples (sum of the diagonal values in the error matrix) to the total reference data. This measure is easy to calculate and enables the quality to be estimated across a number of classes. Quality estimates for individual classes are, however, also possible. The user's and producer's accuracy (Congalton, 1991) is suitable for this purpose. The *producer's accuracy* is calculated from the correctly classified number of objects in a class divided by the number of reference objects in this class. This is a measure of how well the respective class has been identified. It includes the so-called *error of omission*. This corresponds to the reference objects that belong to the class under consideration but were not recognized by the classification model. The *user's accuracy* can be derived from the correctly classified objects of a class divided by the total number of all the objects assigned to this class. This permits a statement to be made on the reliability of the classification for the class under consideration, since it also includes in the measure of quality those objects that do not belong to the class under consideration but which are identified as such objects by the classification model. These objects correspond to the *error of commission*.

Since Research Centre Jülich was selected as the validation area and since precise ground information is available about the Centre, all the objects present can be used as reference objects. As already mentioned, this excludes those objects that contain the segmentation of mixed use and thus cannot be definitely assigned to any class. The statistical information of such a quality assessment is thus 100% representative. An approach enabling the statistical information value of quality assessments to be calculated even with a smaller number of reference objects or in unknown areas is presented in Section 7.1.3.

Table 5.3 summarizes the measures of quality assessment. A total of 2036 reference objects were used to determine the quality. The overall accuracy for the classification set up using SEaTH is 95%. The producer's accuracy is 90% or better for all classes. An exception here is the class **streets**, which achieves 81% in the producer's accuracy. It thus becomes apparent that in almost all classes more than 90% of the reference objects are identified, and in the **forest** class it is actually 98%. The supposed poorer performance of **streets** was already found in the SEaTH analysis where the separability of **streets** from **buildings** was 1.69. The reliability of the classification, represented by the user's accuracy, can also be regarded as very good. For most classes, with the exception of **fields** and **shadow**, the reliability is over 90%. This was also already indicated by the SEaTH analysis, since the separability of the classes of **forest** from **shadow** was only 1.41.

Table 5.3 Quality assessment of the object-based classification of FZJ

	BUILDINGS	FOREST	STREETS	MEADOWS	FIELDS	SHADOW	Sum
BUILDINGS	**188**	0	7	0	0	2	197
FOREST	2	**1393**	0	5	0	5	1405
STREETS	7	0	**84**	0	0	0	91
MEADOWS	4	3	7	**144**	0	1	159
FIELDS	0	0	2	6	**46**	0	54
SHADOW	1	26	0	0	0	**87**	114
UNCLASS	2	1	3	5	5	0	16
Sum	204	1423	103	160	51	95	2036
Producer Accuracy	0.92	0.98	0.81	0.90	0.90	0.91	
User Accuracy	0.95	0.98	0.92	0.91	0.85	0.76	
Overall Accuracy	0.95						

Altogether, the classification result based on the SEaTH analysis can be described as very good. This is supported by the overall accuracy of 95%. Furthermore, SEaTH is also suitable for identifying **buildings** very precisely and reliably. **Buildings** are of special significance since they can later be used to verify the NPT, for example by identifying undeclared buildings or construction work. SEaTH is thus an appropriate analysis tool for rapidly determining good and reliable features for an object-based classification. Furthermore, using SEaTH it is possible to make a quantitative assessment of the features found and thus to employ the calculated separability as a measure of quality for features. In this way, even before a classification has been made, it is possible to make a first assessment of the expected classification result and to compare features with each other.

Chapter 6
Preprocessing of Very-High-Resolution RS Data

Before work can begin on object-based image analysis, the RS data sets must first be appropriately preprocessed. In view of comprehensive automated NPT verification, only those preprocessing steps will be applied that can be automated and which are necessary for further processing.

In this section, new methods developed as part of project work at Research Centre Jülich and in cooperation with the GMOSS expert network will be presented, examined and evaluated with respect to their applicability to very high-resolution RS data.

6.1 Geometric Correction

In order to compare different multitemporal satellite scenes with the aim of change detection, it must be ensured that the pixels are in exact agreement in the individual scenes. The QuickBird-2 data used in this study are already geocoded to a geodetic reference system, the *Universal Transverse Mercator* (UTM, Zone North 34, WGS 84). However, geocoding alone does not guarantee a precise agreement of the image position of the individual pixels in different image scenes since the recording parameters, such as camera angle, recording angle or a slightly different orbit of the satellite, may change in multitemporal scenes.

Part of the present work is concerned with combined change detection at the *Nuclear Fuel Research and Production Centre* (NFRPC) Esfahan. A careful **image to image** registration of the RS data sets must first be undertaken before beginning any other work. To this end, a reference data set (July 2002) was selected and the data set from July 2003 was then geometrically corrected to agree with it. The data set from May 2004 was geometrically corrected to the satellite scene of July 2003. The aim of this step is to achieve as precise an agreement as possible between all the pixels in the multitemporal scenes studied. After a brief basic explanation of the registration routine, the actual registration of the image data is performed by an automated procedure.

Starting from two systems of Cartesian coordinates (see also Fig. 6.1 (A)) describing the arrangement of the pixels in the reference image (x, y) and in the image

© Springer Science+Business Media B.V. 2008

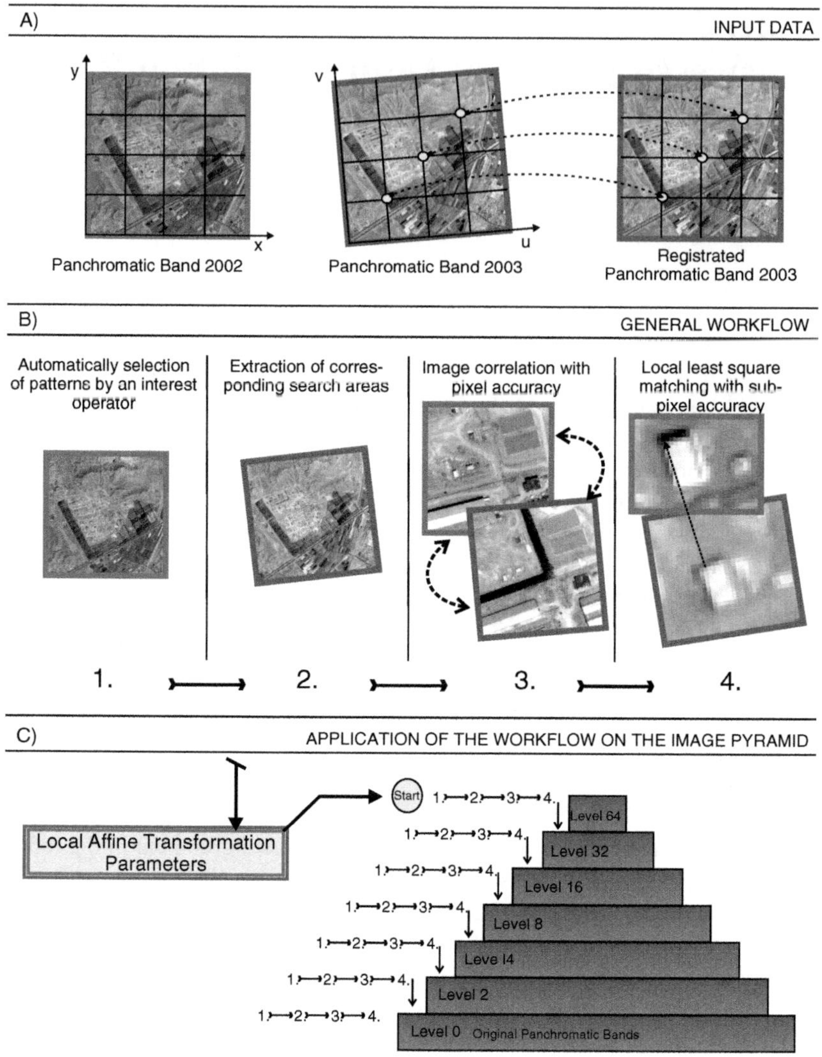

Fig. 6.1 Semi-automatic image registration – image to image

to be registered (u, v), these systems of coordinates can be linked via two functions.

$$u = f(x, y) \qquad v = g(x, y) \tag{6.1}$$

Where f and g are polynomial transformation equations, usually with polynomials of the first to third order (Richards and Jia, 1999).

$$u = a_0 + a_1 x + a_2 y \quad v = b_0 + b_1 x + b_2 y \tag{6.2}$$

Coefficients a_i and b_i of the first order transformation represented in Equation (6.2) are determined by tie points (TP) or ground control points (GCPs). This may be points in the terrain such as road junctions or field boundaries. For first-order polynonomials it is sufficient to unambiguously identify three TP while third-order polynomials require at least ten. The *least squares method* is then used to establish geometric relations between the two images and calculated the coefficients. The position of the pixels in the image to be registered is now recalculated with the aid of the transformation equation so that these pixels correspond to the image position in the reference image. The actual transformation can be performed by various resampling methods such as *nearest neighbour*, *bilinear interpolation*, *cubic convolution* or *sin(x)/x*. The above-mentioned methods are well known in remote sensing and are discussed in detail in the standard literature such as Richards and Jia (1999) or Schowengerdt (1997).

At the moment, manual registration of the image data is a widely used technique in remote sensing where the user has to manually determine the tie points in both images. However, depending on the size of the satellite images to be processed this method is very tedious and time-consuming. Furthermore, due to the continuously increasing resolving power of satellite image sensors it may be very difficult or even impossible to unambiguously identify TPs.

An example may be the junction of a major road (15 m wide). In an ASTER satellite image with a spatial resolution of 15 m, this junction corresponds to precisely one pixel. With the very high-resolution QuickBird data used in the present thesis (resolution 0.6 m), this junction would consist of 625 pixels. It is very difficult, and also labour-intensive and time-consuming to precisely determine the corresponding pixel in comparison to the reference image by hand (and eye). Manual, visual image registration clearly comes up against its limits in view of such a high spatial resolution.

With respect to automating applications and the limits discussed above, automatic, computer-controlled methods are becoming more and more important. In this connection, attention is drawn to work by Dai *et al.* (1996), Dai and Khorram (1997), Growe and Tönjes (1997), Lee *et al.* (1993), LeMoigne (1998), Li *et al.* (1995) and Thepaut *et al.* (1998). These authors indicate the various possibilities and approaches for automating the registration process.

The present study uses a semi-automatic registration routine developed by the *German Aerospace Centre (DLR)* in Oberpfaffenhofen. The cooperation and transfer of know-how is being implemented as part of the EU project Global Monitoring for Security and Stability (GMOSS) (see Section 1.3).

6.1.1 Semi-Automatic Registration Routine

The registration routine developed at the DLR in Oberpfaffenhofen was originally designed for the application of 3-line stereo images. Due to continual further developments, it is, however, also possible to register multitemporal satellite images which are also of a stereo character. This field of application will be investigated in the following in cooperation with DLR. Basic data are the QuickBird satellite images of the NFRPC Esfahan from 2002, 2003 and 2004. The temporal difference in the images amounts to about one year in each case. This time span, which can certainly be regarded as very long, makes great demands on automation software since both many of the natural environmental parameters and also the parameters for taking the images have changed. The aim of the software is the precise identification of a large number of TPs, on the basis of which the data will ultimately be registered.

The software basically operates in four phases, which will be briefly described. Figure 6.1(B) additionally shows a schematic of the individual steps. For more detailed information on the operating mode of this routine see Lehner and Gill (1992, 1989) and Lehner (1986).

The first step is the automatic search for patterns in the reference image that are suitable for registration. A so-called *interest operator*, developed from the well-known *Förstner operator* (Förstner and Gülch, 1987) selects these patterns. In doing so, the interest operator automatically decides whether a pattern is suitable or not. A weighted centroid is calculated according to the least squares method for each of these potential patterns. This centroid is then weighted by the associated first derivation in the column and line direction. The two parameters of *roundness* and *size* are extracted from the error ellipsis of this fit. A pattern and the associated centroid are only suitable as a tie point if the parameters of size and roundness are below a given threshold.

If the patterns are determined in the reference image then corresponding search windows are identified in other scenes. In these windows, the algorithm searches for the patterns previously determined in the reference image. The search windows are determined on the basis of a so-called *local affinity transformation* of the image pairs. The parameters required for the affinity transformation can be calculated by the least squares method on the basis of known TPs. These known TPs are determined manually in the top level of the image pyramid before the start of the actual registration routine. As a rule, four TPs are sufficient for this purpose. These clearly corresponding points are easier to determine on the top level of the image pyramid since the image resolution and thus the parallaxes are accordingly reduced there. It is only necessary to select the TPs manually once and this serves exclusively to determine the affinity parameters. In order to avoid serious misallocations of the patterns, only those search windows are accepted whose RMS[1] errors are below a given threshold. The number and distribution of the search windows that are not accepted furthermore point to local areas in which automatic tie point determination is difficult.

A correlation matrix is now calculated for a given pattern with the associated search window. The maximum of the correlation coefficient is defined by the pixel in the search window that corresponds to the centroid pixel of the reference pattern.

[1] Root mean square error.

A quality estimate and acceptance rules then finally determine whether the identified pixels can be used as TPs.

In a final step, the results are improved in the subpixel range by the iterative application of the *local least squares matching* technique.

The steps are applied on each level of the image pyramid (cf. Fig. 6.1(C)). The pyramid is created by calculating down the spatial pixel resolution. In doing so, the resolution is enlarged by a factor of two in each of the six steps. The advantages of the image pyramid are a reduction of the image parallaxes and a simplified identification of the TPs for the affinity transformation. Furthermore, the quality of the final results benefits very greatly from the fact the in an image pyramid the differences in disturbances and shifts of the satellite images under consideration are relatively small from one level to the next.

Figure 6.2 shows the TPs found by the routine for the panchromatic satellite image of July 2003. An overview of the results of this registration is given in Table 6.1. It can be clearly seen that with increasing resolution the number of patterns found by the interest operator also grows. Of the total number of patterns found, about 56%–68% are classified as "suitable for registration" in all levels of the image pyramid. In the case of the pan image used with an image size of 5300 × 5800 pixels, no less than 52 000 TPs are possible. After applying the steps (two–four) described above, only 1801 possible tie points remain. This results from the application of different threshold values and acceptance rules in finding the search window and in

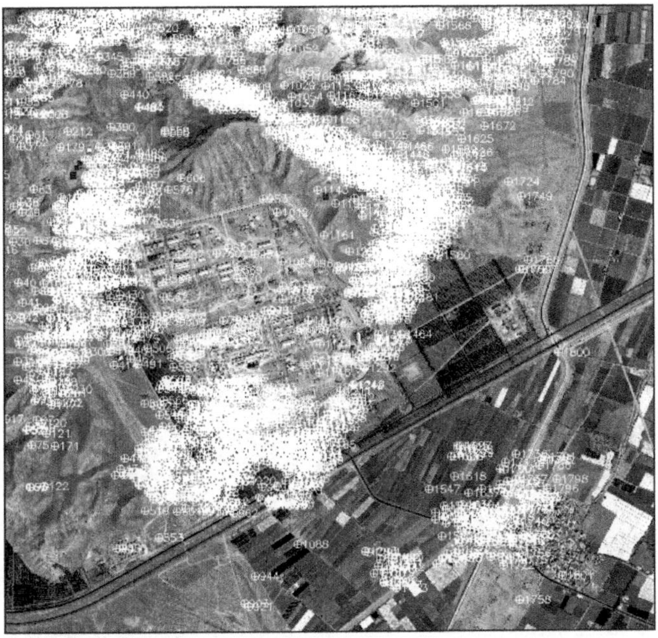

Fig. 6.2 Automatically identified TPs for registering the image data of July 2003 on July 2002 (reference)

Table 6.1 Results of the automatic registration routine

Image Pyramid	Intr.-op. Points	Possible TPs total (%)	Computing time (s)
Level 64	56	38 (67.9)	1
Level 32	210	135 (64.3)	2
Level 16	735	453 (61.6)	4
Level 8	2378	1446 (60.8)	6
Level 4	7592	4512 (61.8)	25
Level 2	26132	14672 (56.1)	140
Original Image	91208	52474 (57.5)	900
Full Process	Image-size	Total TPs* accepted	Total time
	5300 × 58000	1801	60 min

*After the application of threshold criteria, acceptance rules and TP grid processing

the correlation. In this way, only the best TPs are used. The points are then plotted in a grid with a mesh width of 100 × 100 pixels. The total time for the semi-automatic registration of this pan scene amounts to about 60 min. This amount of time can be regarded as fast in view of the 1801 points used and the good quality of the final result.

The example of the semi-automatic registration of the July 2003 scene on the July 2002 reference image shows that even with a high temporal difference a very large number of points (1801 TPs) are still found. Figure 6.3 shows the result of the transformation of the pan image of July 2003 with a first-order polynomial and a bilinear resampling method. A section from the recorded image from 2003 is superimposed on the reference image. It can thus be seen that the positions of the streets and the edges of the two images are now in good agreement.

Fig. 6.3 Results of the registering routine of the panchromatic image data of July 2003 on July 2002 (reference). The 2003 scene is projected onto that of 2002. Original Data/Product © Quickbird/Digital Globe; 2003; Distributed by Eurimage

A further step towards a geometric correction would be orthorectification. This evens out topographical or height-related inaccuracies in the satellite scene. The basis of such an orthorectification is usually a *digital elevation model* (DEM). In the case of the very-high-resolution QuickBird-2 RS data used, a very-high-resolution digital elevation model, or even better a *digital surface model* (DSM), would be required for orthorectification. With a very good and very high-resolution digital surface model, it would then also be possible to even out the positional inaccuracies of high buildings which result from the different camera angles of the satellite. Unfortunately, no such DEM/DSM is currently available for the regions studied. Attempts at extracting a very high-resolution DSM from stereo satellite images are being made at present within the GMOSS network. However, the results obtained so far are not yet suitable for orthorectification. The DEM obtained from Aster satellite data cannot be used for this study either due to the spatial resolution of 15 m at maximum, which is too low. Due to the lack of a suitable basis for orthorectification, it had to be dispensed with in this study. However, since the nuclear facilities studied are all situated within a flat region, the lack of orthorectification does not have a serious effect except for large differences in camera angle and very high buildings.

6.2 Pan Sharpening – A Comparison of Methods

The satellite data used in the present study include both panchromatic and also multispectral data. The panchromatic data have a very high spatial resolution and a low spectral resolution. The case is exactly the opposite with multispectral data.[2] A high spatial resolution is necessary in order to identify shapes, edges and structures accurately. Different types of land use and objects can, on the other hand, be identified by their spectral properties. In order to use high-resolution *and* multispectral information in one data set, it is necessary to combine the data of high spatial resolution with those of high spectral resolution.

This combination is termed *pan sharpening* and is one of the group of data fusion techniques (after: EARSeL)) (Ranchin *et al.*, 2003).[3] Several algorithms with various features are available for the field of remote sensing. As part of this study, some of these methods will be briefly presented and assessed with respect to their results. The technologies used here are especially suited for the pan sharpening of high-resolution satellite images. The algorithms investigated are:

- *Wavelet Fusion – Daubeschies Wavelet*
- *A-Trous – Cubic Spline Filter*
- *PC Spectral Sharpening*
- *Gram–Schmidt Spectral Sharpening*

---

[2] The attributes low and high relate to the satellite data used. In comparison, for example, to Landsat TM data the spatial resolution of the multispectral data used is undoubtedly to be regarded as very high.

[3] Since in this chapter consideration will be given exclusively to the data fusion method of pan sharpening the terms data fusion and pan sharpening can be used synonymously.

The *wavelet fusion* and the *cubic spline filter* methods were programmed in IDL and implemented in the ENVI satellite image software used here. The *PC spectral sharpening* and *Gram–Schmidt spectral sharpening* methods are internal algorithms in ENVI for image fusion.

A good result is achieved from image fusion when the resulting image is as realistic as possible. The fused image should thus represent what would be seen if the multispectral sensor had the spatial resolution of the panchromatic sensor (Ranchin and Wald, 2000). This means an improved basis for interpretation with constant **spectral authenticity** of the results. In this connection the concept of spectral authenticity means that the original spectral values of the multispectral data are modified as little as possible. This point is of special significance since although older algorithms, such as the IHS method, visually sharpen the image very well, nevertheless they modify the original spectral values. This is usually sufficient for the field of photo interpretation or visual interpretation, but in many cases it is not sufficient for the field of remote sensing on the basis of computer-controlled interpretation steps.

The basis of the comparison is a QuickBird-2 satellite image of NFRPC Esfahan of 7 July 2003. Figure 6.4 shows the spatially highly resolved panchromatic channel for a subset of this scene and the spectrally highly resolved multispectral channels in a colour combination (B4, B3 und B2, magnified fourfold). Different sections of the overall satellite image are used for comparing results and methods. These sections vary depending on the extent to which the area is built on and whether vegetation is present (cf. Fig. 6.5).

After a brief discussion of the data fusion methods tested, the results will then be compared.

6.2.1 Wavelet Fusion – Daubechies Wavelet

Simplified Wavelets are something to the effect that mathematical functions which can, for example, be used to compress data. **Daubechies wavelet** discussed here

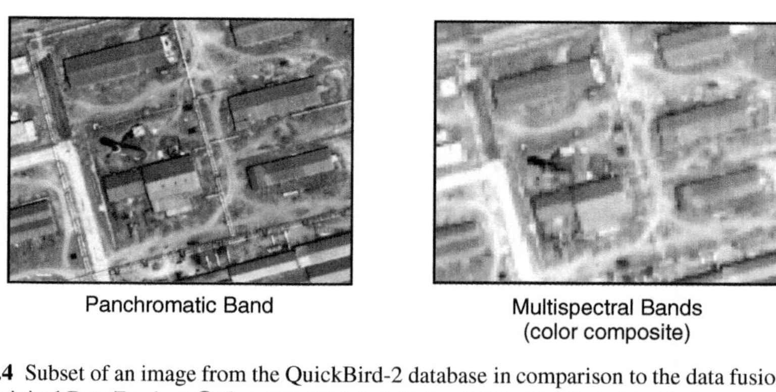

Fig. 6.4 Subset of an image from the QuickBird-2 database in comparison to the data fusion methods. Original Data/Product © Quickbird/Digital Globe; 2003; Distributed by Eurimage

Fig. 6.5 The test areas facility, dense vegetation and little vegetation. Original Data/Product © Quickbird/Digital Globe; 2003; Distributed by Eurimage

compresses the original image via a filter bank and thus extracts the image information, the so-called *wavelet coefficients* (WCs). These WCs contain the *low-frequency* and *high-frequency* information of the initial image, i.e. details and spatial information. Using this information it is possible to recreate the initial image without any loss via an inverse wavelet transformation. (Ranchin *et al.*, 2003; Ranchin and Wald, 2000)

The wavelet transformation takes place on the basis of the wavelet filter bank (WFB) shown in Fig. 6.6 since the wavelet transformation of two-dimensional image data corresponds to an iterative arrangement of high- and low-pass filters (Mallat, 1989).

The first application of this WFB compresses the original panchromatic channel (pan) by a factor of 2 and saves the associated wavelet coefficients WC^H, WC^V, WC^D[4] (cf. Fig. 6.7). Another iterative application of the filter bank leads to a

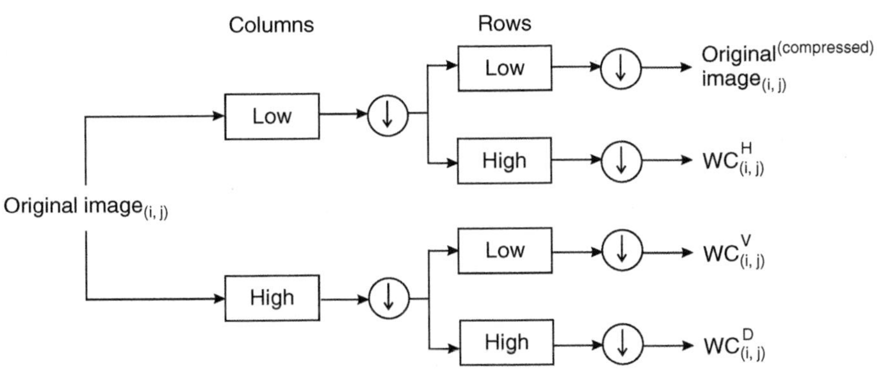

Fig. 6.6 Wavelet filter bank of a Daubechies wavelet. Low corresponds to a low-pass filter and High to a high-pass filter

[4] This corresponds to the information of the initial image in the horizontal (H), vertical (V) and diagonal (D) direction.

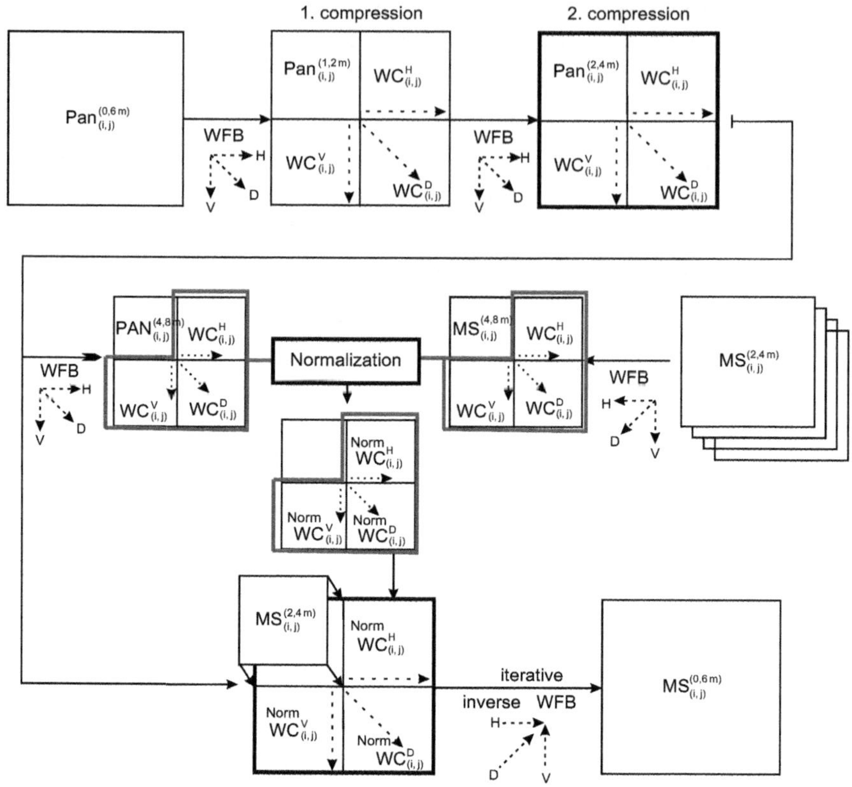

Fig. 6.7 Application of Daubechies wavelet method to a QuickBird-2 satellite image. With $Pan_{(i,j)}$ = panchromatic channel, $MS_{(i,j)}$ = multispectral channel, WFB = wavelet filter bank and $WC_{(i,j)}$ = wavelet coefficients

panchromatic band compressed by a factor of 4. With the QuickBird data this results in a spatial resolution of 2.4 m, which corresponds to the resolution of the multispectral channels (MS). The subsequent fusion procedure then takes place separately for each MS channel.

A necessary step for fusing the data is a normalization of the wavelet coefficients. To this end, an MS channel and the compressed pan channel are transformed again by the WFB. In the first step the normalization parameters are calculated for the resulting wavelet coefficients (WC^z, $z = H, V, D$):

$$a^z = \sigma^z_{ms} / \sigma^z_{pan} \tag{6.3}$$

$$b^z = m^z_{ms} - a^z m^z_{pan} \tag{6.4}$$

with σ = standard deviation and $m = mean$
These parameters are used to normalized the wavelet coefficients of the panchromatic channel to those of the multispectral channel.

$$WC^{z}_{(i,j)} \rightarrow a^{z} WC^{z}_{(i,j)} + b^{z}, \quad z = H, V, D \tag{6.5}$$

The panchromatic image compressed to 2.4 m is now replaced by the 2.4 m multispectral image and using the normalized coefficient (WC^{Norm}) is reconstructed to a spatial resolution of 0.6 m via an inverse application of the wavelet filter bank. The normalization and substitution are performed separately for each multispectral channel. The result is a multispectral data set with an increased spatial resolution (Ranchin *et al.*, 2003; Ranchin and Wald, 2000).

6.2.2 A-Trous – Cubic Spline Filter

In contrast to Daubechies wavelet transformation discussed above, the **cubic spline filter (CSF)** does not compress the initial image but rather "blurs" the data somewhat. The details can be extracted by determining the difference between the initial imaged and the blurred imaged. If in an inverse transformation the details are added to the blurred image then the initial image can be reconstructed without loss.

The blurring results in a deterioration of the resolution by a factor of 2. The filter must be stretched in order to further blur the initial image and thus increase the resolution even more. This can be done by adding a $'0'$ (à trous = a hole) between each filter coefficient. The set-up of the filter is shown in Fig. 6.8. More detailed information can be found in Aiazzi *et al.* (2002).

In a first step for pan sharpening, the CSF is applied to the pan band (A) (cf. Fig. 6.8). The details $P-1$ result from the difference of the pan band (A) and the result (A'). The result of the first transformation (A') is now transformed by the à-trous CSF to (A''). The difference of $A' - A''$ yields the details $P-2$. In the same way, a transformation is now performed with a multispectral band X. However, prior to this operation it is extrapolated to the resolution of the pan band. Figure 6.8 shows the spatial dimension of the extrapolated MS band X reduced fourfold. This does not influence the transformation but is done for reasons of clarity. The transformation yields two blurred MS bands X' and X'' as well as the details $MS-1$ and $MS-2$. The details of the pan channel $P-1$ and $P-2$ and the MS channel $MS-1$ and $MS-2$ are then normalized to each other. A similar normalization routine is used to that applied for the Daubechies wavelet transformation already discussed. The MS channel is sharpened via an inverse transformation. The twice-filtered MS channel X'' is first added to the normalized details $P-2$. By a repeated addition of the result to the normalized details $P-1$ a spatially highly resolved multispectral channel is obtained. This sharpening procedure is applied separately for each MS channel.

6.2.3 PC Spectral Sharpening

PC stands for *principal components*. Principal components are synthetic channels generated by a rotation within the feature space. The aim of the principal component

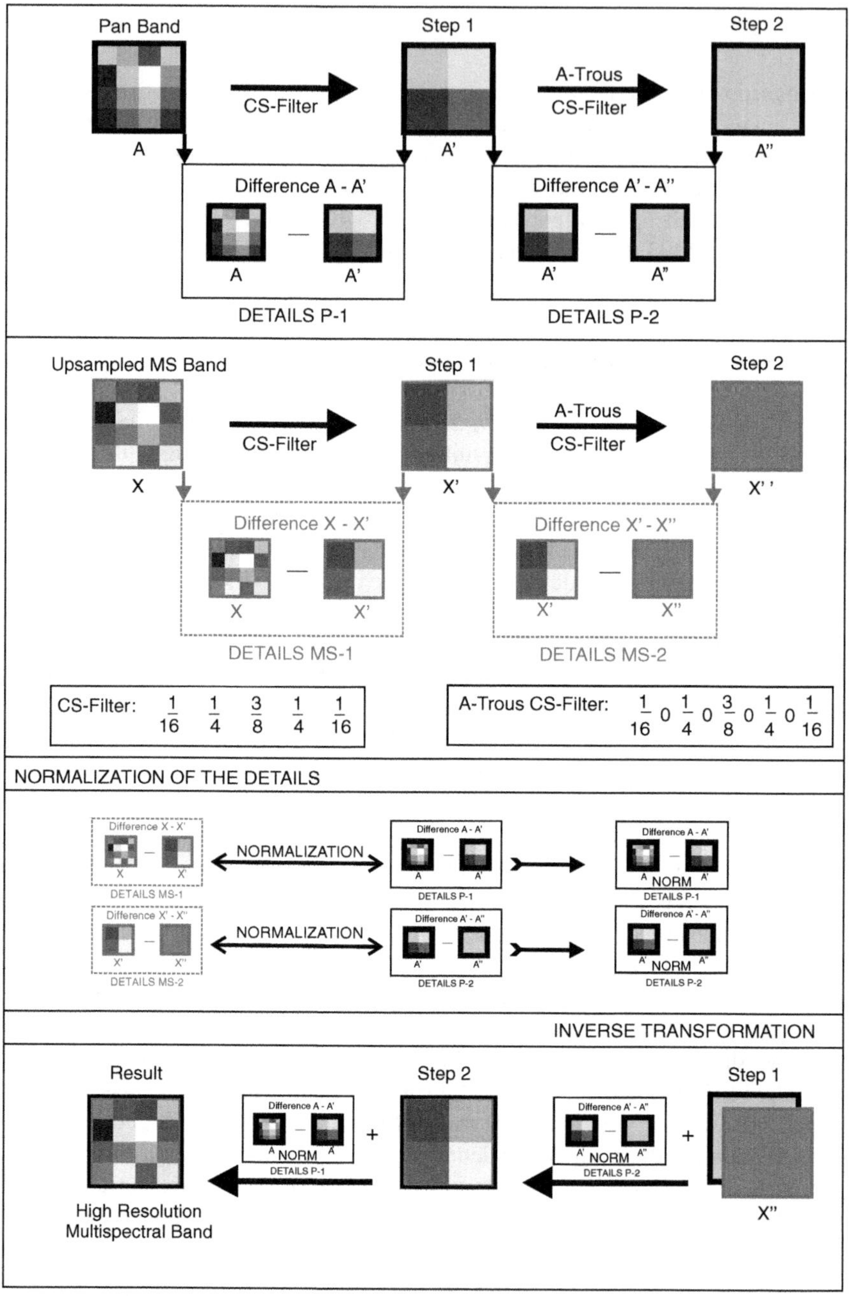

Fig. 6.8 Pan sharpening with à trous – cubic spline filter

transformation is to filter out redundant information in the original satellite channels. By a rotation and shift of the original coordinate systems in bidimensional feature space the principal components are concentrated along the maximum variance. As a rule, the first three principal components already contain 95% of the total variance. A high proportion of the information is concentrated in the first principal component.

The principal component transformation is a method frequently used for information concentration in remote sensing. This has been discussed in detail on many occasions in the literature so that it will not be described further here. Attention is instead drawn to the existing body of literature such as Schowengerdt (1997) or Mather (2004).

Via a principal component transformation, the **PC spectral sharpening method** (cf. Fig. 6.9) generates the four principal components from the multispectral QuickBird dataset. The first principal component, which contains the major fraction of the total variance, is then fused with the spatially highly resolved panchromatic band by a fitting procedure. An inverse principal component transformation is then performed with the fused first principal component and the three remaining principal components. An automatic resampling of the second, third and fourth principal component is performed at the same time in this inverse transformation. The result is a spatially highly resolved multispectral data set.

6.2.4 Gram–Schmidt Spectral Sharpening

The **Gram–Schmidt spectral sharpening** method functions in a similar manner to the PC spectral sharpening discussed above, also with the aim of minimizing redundancy. However, in this case the main component transformation is replaced by a Gram–Schmidt orthogonalization. The Gram–Schmidt transformation is a common method in linear algebra and multivariate statistics. For a more detailed discussion of this methodology see Clayton (1971) and Farebrother (1974).

In the first step, a simulated panchromatic band is generated (cf. Fig. 6.10). This can be constructed from a single multispectral band or from the mean of all the multispectral bands. Since taking the mean provides the best results, this method will be used in the following. The simulated panchromatic band and the original multispectral bands are now subjected to a Gram–Schmidt transformation. The first multispectral band is replaced by the simulated panchromatic band. The Gram–Schmidt components (G-S-Cs) are obtained as a result. A modified pan band of higher spatial resolution is generated by comparing the statistics of the first Gram–Schmidt component (G-S-C-1) with the statistics of the original panchromatic image. The first of the Gram–Schmidt components is then replaced by the modified pan band. The panchromatically sharpened multispectral channels are obtained via an inverse application of the Gram–Schmidt transformation. Further information on the Gram–Schmidt spectral sharpening algorithm can be found in Laben and Brower (2000).

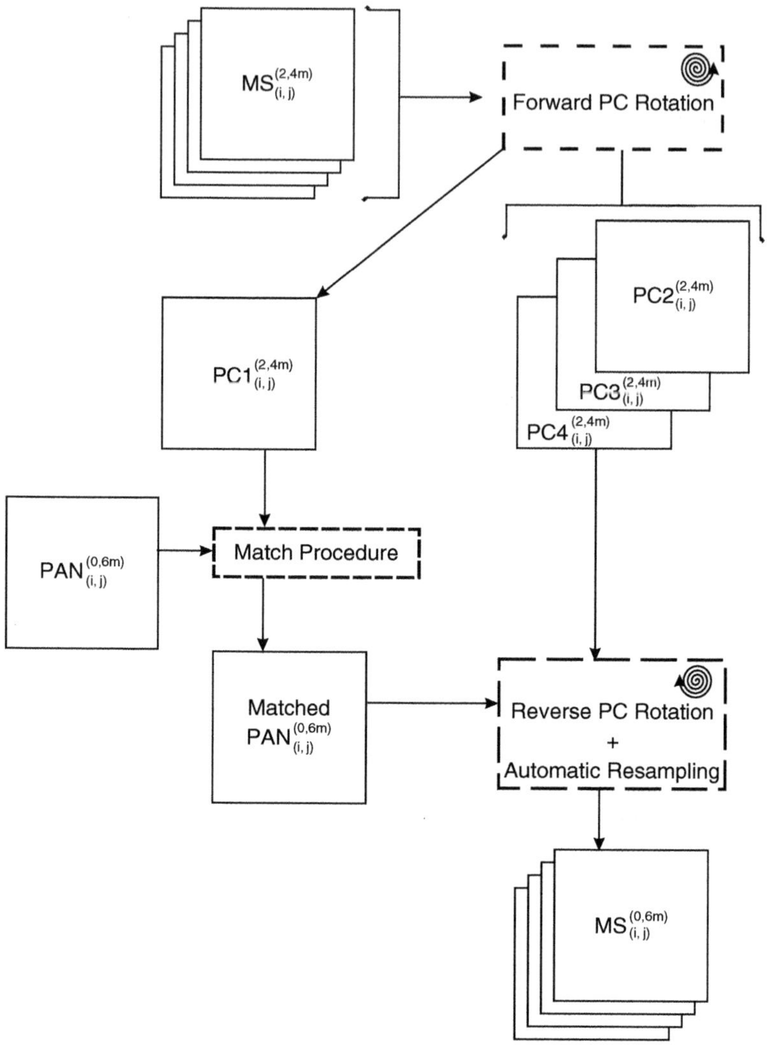

Fig. 6.9 Application of PC spectral sharpening to a QuickBird-2 scene

6.2.5 Qualitative and Quantitative Evaluation of the Results

The selected subsets of the QuickBird-2 satellite image from 2003 were pan sharpened on the basis of the registered satellite data (cf. Section 6.1).

A visual comparison of the results (Fig. 6.11) shows that all the methods significantly improve the interpretation basis for more advanced studies. The image sharpened by Daubechies wavelet makes the poorest visual impression. The edges make a step-like impression and the whole scene appears streaky. The reason for

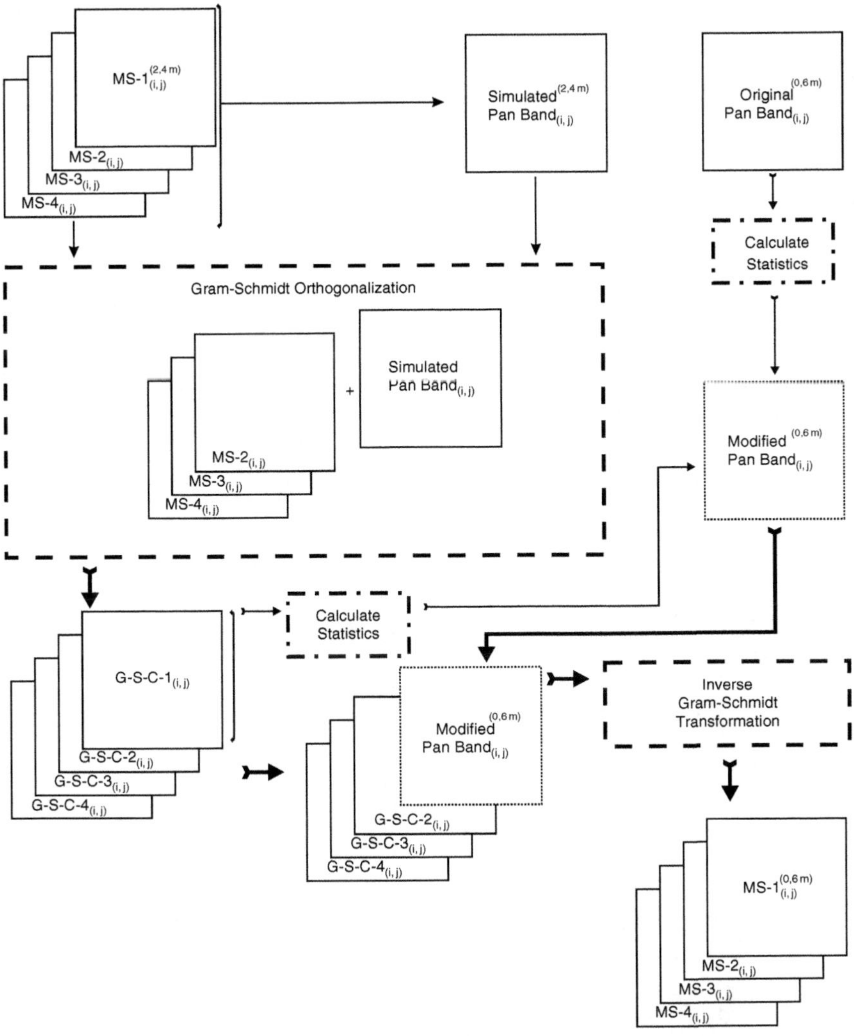

Fig. 6.10 Application of the Gram–Schmidt spectral sharpening method to a QuickBird-2 scene

this is probably that this method incorporates a greater proportion of multispectral data in the final result than the other techniques. Furthermore, slight inaccuracies in registering the initial data are intensified by this method, which can result in a streaky appearance. The other three methods provide a very good visual result. The whole scenes are very sharp and the edges also make an optically good impression.

The colouring of the upper results differs from that of the lower results. The colours of the images sharpened by Daubechies wavelet and à trous CSF are stronger and display greater contrast. The extent to which the spectral authenticity is really

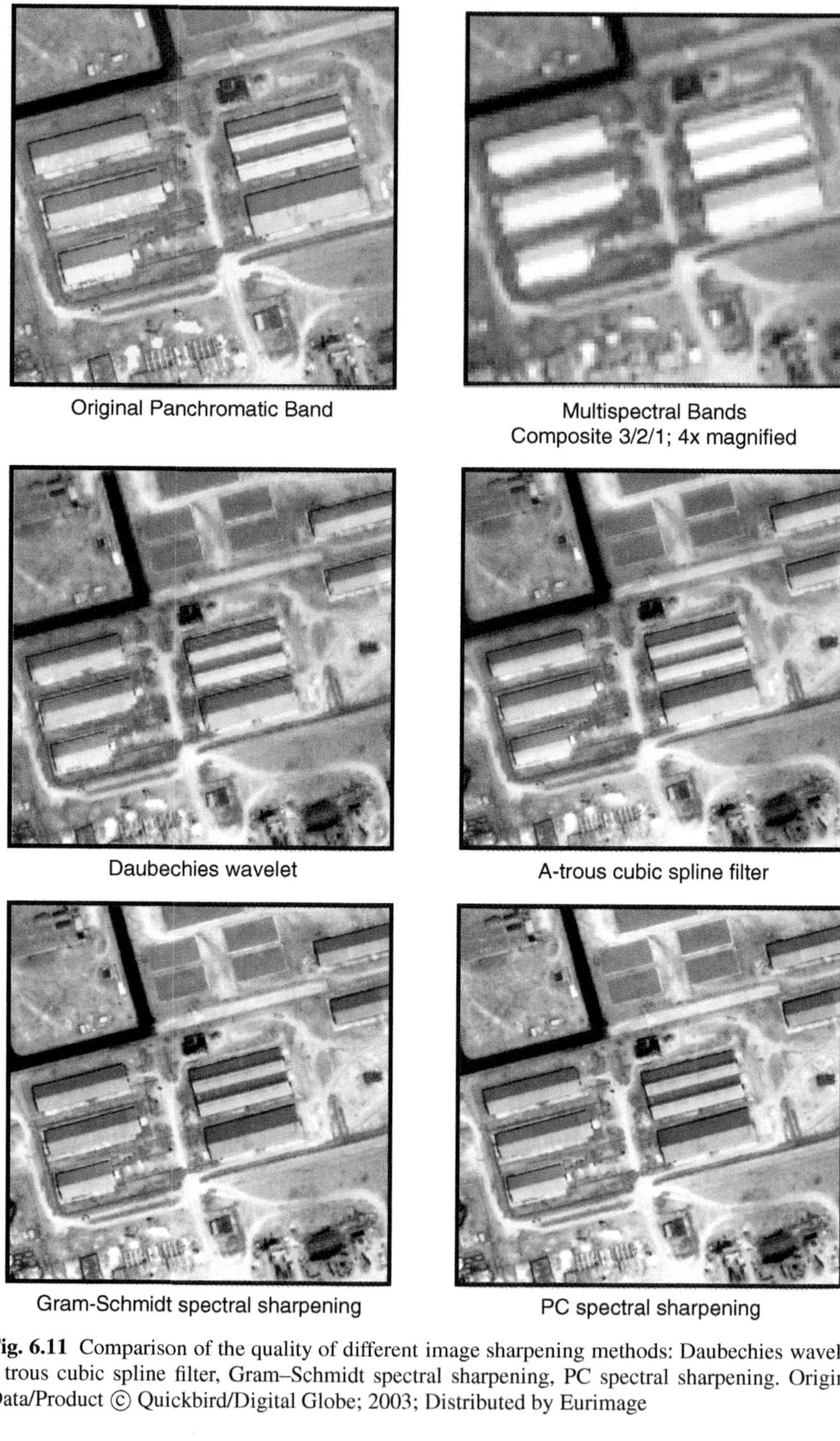

Fig. 6.11 Comparison of the quality of different image sharpening methods: Daubechies wavelet, à trous cubic spline filter, Gram–Schmidt spectral sharpening, PC spectral sharpening. Original Data/Product © Quickbird/Digital Globe; 2003; Distributed by Eurimage

maintained is difficult to judge visually. This aspect will be investigated in the following by the *universal image quality index.*

6.2.5.1 Universal Image Quality Index

There are several methods for determining the quantitative quality or deviation of image data. These methods can be basically divided into two classes. The first class contains mathematically defined measures such as the *mean squared error (mse), the peak signal to noise ratio (psnr), the root mean squared error (rmse)* and the *mean absolute error (mae)*. The second class is composed of methods based on human perception (*human visual system, hvs*) and includes a visual estimation of quality (Pappas and Safranek, 2000).

The universal image quality index developed by Wang and Bovik (2002) can be regarded as one of the mathematical methods. In comparison to the above-mentioned methods it is simple and straightforward. This has the advantage that this index can be readily implemented in the ENVI satellite software used here and that only a modest amount of computing power is required for the calculations (i.e. short computing times).

With the original image $x = [x_i | i = 1, 2, \ldots, N]$ and the test image $y = [y_i | i = 1, 2, \ldots, N]$ the universal image quality index (Q) is defined as follows:

$$Q = \frac{4\sigma_{xy}\bar{x}\bar{y}}{(\sigma_x^2 + \sigma_y^2)[(\bar{x})^2 + (\bar{y})^2]} \tag{6.6}$$

with

$$\bar{x} = \frac{1}{N}\sum_{i=1}^{N} x_i, \qquad \bar{y} = \frac{1}{N}\sum_{i=1}^{N} y_i,$$
$$\sigma_x^2 = \frac{1}{N-1}\sum_{i=1}^{N}(x_i - \bar{x})^2, \qquad \sigma_y^2 = \frac{1}{N-1}\sum_{i=1}^{N}(y_i - \bar{y})^2$$
$$\sigma_{xy} = \frac{1}{N-1}\sum_{i=1}^{N}(x_i - \bar{x})(y_i - \bar{y})$$

The dynamic range of Q is [−1,1]. The best achievable value is consequently 1 whenever $y_i = x_i$, i.e. the original image and the test image are identical. Q can be rewritten as a combination of three factors:

$$Q = \frac{\sigma_{xy}}{\sigma_x\sigma_y} * \frac{2\bar{x}\bar{y}}{(\bar{x}^2) + (\bar{y})^2} * \frac{2\sigma_x\sigma_y}{\sigma_x^2 + \sigma_y^2} \tag{6.7}$$

The first factor in the equation gives the correlation coefficient of x and y. This factor measures the degree of linear agreement and in the ideal case[5] is thus = 1 and if there is no correlation then = 0. The second factor compares the means of the two images. The range of values is between 0 and 1. The third factor finally examines the variance of the two images. In this case the dynamic range is also [0, 1]. In

[5] The two images are identical.

conformity with the correlation, mean value and variance, these three factors provide a value of 1 and thus Q is also equal to 1.

In the first step, the pan-sharpened test image is calculated down to the resolution of the reference image. This procedure is called downsampling. The individual bands of the reference and test image are then compared with the *universal image quality index*. The results of this comparison are listed in Table 6.2.

A comparison of the different image fusion methods is performed for different subsets of the satellite image. The first has hardly any vegetation (**little vegetation**), the second consists mainly of vegetation (**dense vegetation**), and the third shows a section of the NFRPC Esfahan (**facility**). In this comparison of methods, special attention is paid to the facility area since this is the major focus of the present study. The choice of different sections is, however, intended to reveal possible trends in the quality of the image fusion methods which could occur due to different types of ground cover.

Table 6.2 shows that with respect to the spectral authenticity of the results, Daubechies wavelet consistently displays the best results ($Q = 0.9+$), followed

Table 6.2 Comparison of the quality of different image diffusion methods. Results of the universal image quality index Q

	Facility Q	Dense Vegetation Q	Little Vegetation Q
Daubechies Wavelet			
Band 1	0.93	0.90	0.88
Band 2	0.94	0.90	0.88
Band 3	0.94	0.90	0.89
Band 4	0.94	0.90	0.89
A-Trous Cubic Spline Filter			
Band 1	0.81	0.75	0.71
Band 2	0.83	0.77	0.73
Band 3	0.84	0.77	0.74
Band 4	0.84	0.78	0.74
Gram–Schmidt Spectral Sharpening			
Band 1	0.72	0.63	0.64
Band 2	0.72	0.68	0.59
Band 3	0.75	0.66	0.56
Band 4	0.81	0.83	0.60
PC Spectral Sharpening			
Band 1	0.73	0.43	0.65
Band 2	0.73	0.50	0.60
Band 3	0.75	0.47	0.57
Band 4	0.81	0.84	0.61

by the à trous cubic spline filter with values of $Q = 0.71$–0.84. The Gram-Schmidt and PC spectral sharpening algorithms achieve on average quality values of about 0.6–0.7. If the individual sections are considered in a differentiated manner, then the best spectral authenticity is found in the **facility** area for all the techniques tested. The lowest quality values were achieved in the area of **little vegetation**. The exception here is the PC spectral sharpening algorithm, which yields very poor Q values (0.43+) in the field of **dense vegetation**.

By way of summary, it can be said that there are significant differences in the quality of the sharpening methods with respect to the maintenance of spectral values. These vary in the **facility** area by up to 0.22 and achieve their maximum difference of 0.5 in the **dense vegetation** image segment. With a maximum index value of $Q = 1$, this represents a deterioration. With respect to the image content tested (facility, dense vegetation, little vegetation), it was not possible to identify any significant trends in the quality of the sharpening results.

In the case of the data used in this study, it can be assumed that Daubechies wavelet provides the best results of spectral authenticity. However, due to its poor performance in the visual test this method will not be taken into consideration for further use. The à trous cubic spline filter represents a good alternative for the sharpening of satellite data. In the quality estimate given above, this method provides values ranging from $Q = 0.81$ to 0.85. These Q values are therefore only about 0.1 lower than those of the Daubechies wavelet. It is, moreover, the method which makes the sharpest optical impression. It is also decisive that the edges of the buildings are retained thus making it easier to identify buildings as separate objects. Consequently, the QuickBird-2 satellite images sharpened by the à trous cubic spline filter were selected as the data basis for more detailed studies.

6.3 Radiometric Normalization

With multitemporal satellite images, the external parameters vary when the image was recorded, such as position of the sun, transmission properties of the atmosphere and sensor sensitivity. In order to ensure comparability of the satellite images, the changes in reflectance due to conditions prevailing at the time the image was recorded must usually be corrected before an analysis is made. This is done by a radiometric normalization of the satellite data.

An absolute radiometric correction can be made on the basis of complex atmospheric models such as 6S Code (Vermote *et al.*, 1997), MODTRAN (Berk *et al.*, 1989) or LOWTRAN (Kneizys *et al.*, 1988; Richter, 1990; Sandmeier, 1997). These models take into consideration the position of the sun, the sensor calibration and the properties of the atmosphere at the time the image was recorded. Since these models are only available for certain satellite data, and are also very time-consuming and expensive, a relative radiometric normalization was applied in the present study.

A relative radiometric normalization is based on the inherent radiometric information of the satellite data itself. It normalizes the reflectances of all satellite images used to that of a reference scene. Various methods are available for the relative radiometric normalization of multitemporal satellite data such as those of Schott *et al.* (1988), Hall *et al.* (1991), Moran *et al.* (1992), Furby and Campell (2001) and Du *et al.* (2002).

All the methods are based on the assumption that the different reflectances of unchanged image regions in the multitemporal data can be approximated via linear transformations. The difficulty consists in identifying these time-invariant image regions. With their MAD alteration technique, Nielsen *et al.* (1998) and Nielsen *et al.* (2002) provide a good opportunity of identifying the time-invariant image regions. These thus correspond to the *no change pixel* in a satellite image. As described in Section 4.2, the MAD methodology is invariant to linear changes, i.e. a relative radiometric normalization would not be necessary for alteration detection on the basis of MADs. Since in the present study object-based classifications were performed and the classification model obtained is transferred to other satellite images as a reference model, a radiometric normalization is, however, necessary.

In the first step, the MAD components are calculated for radiometric normalization. Once the MADs have been obtained in the scenes under consideration, then a decision must be taken on identifying the *no change pixels*. All the pixels fulfilling the following condition are used for radiometric normalization:

$$\sum_{i=1}^{N}\left(\frac{MAD_i}{\sigma_{MAD_i}}\right)^2 < S \tag{6.8}$$

i corresponds to the number of MADs[6] and S is a decision threshold. Assuming time invariance, the sum of the standardized MADs shown above has an approximately chi square distribution with N degrees of freedom. The threshold value S is laid down as $S = \chi^2_{N,W=0.01}$ with W as a probability of S or less. The pixels thus identified represent the time-invariant image regions as long as the totality of the radiometric differences of the two satellite images under consideration are based on linear effects.

The data are usually normalized on the basis of the no change pixels by means of an *ordinary least squares (OLS)* regression analysis (Freund, 2003), (Yang and Lo, 2000). However, this only permits uncertainties in one of the two variables (dependent and independent variables). In a normalization both variables have uncertainties since the determination of a reference scene and a scene to be normalized is arbitrary. An alternative is *orthogonal linear regression (OLR)*, which handles both sets of data symmetrically. In the present study, OLR is used for normalization. Derivation and a discussion of this regression analysis can be found in Canty *et al.* (2004).

---

[6] There are 4 MAD components with the QuickBird data.

Another advantage of radiometric normalization based on MADs is that it can be automated. The procedure is implemented as an open source IDL program. After selecting the satellite images to be normalized, the remaining procedure is completely automatic. It should be noted that the satellite scenes must be registered relative to each other since a precise agreement of the pixels in the two scenes is the basis for the MAD technique.

Chapter 7
Object-Based Analysis of Iranian Nuclear Facilities

This chapter deals with the development, application and assessment of a methodology for (semi-)automated, object-based image analysis in order to make available specific information for NPT verification from very highly resolved RS data. The methodology is demonstrated with the example of RS data on nuclear facilities in Iran. The methodology present comprises all the steps necessary within object-based image analysis such as preprocessing, segmentation, feature analysis, semantic modelling, classification and evaluation of the results.

In the further course of this thesis, the extraction of typical features for the object classes of nuclear facilities and the compilation of reference models for object-based classification will be discussed, and their temporal and spatial transferability will be examined and assessed.

7.1 Site – NFRPC Esfahan, 2002

The *Nuclear Fuel Research and Production Centre* (NFRPC) Esfahan is regarded as a primary centre of the Iranian nuclear programme and is Iran's largest nuclear research centre. It was founded in 1974 to the south east of the city of Esfahan. A brief description of the facts known about the area under investigation can be found in Section 1.2. In general, there is little publicly accessible information about NFRPC, so that there is, for example, no precise description of the purposes of individual buildings or parts of the facility. The following analyses are therefore based on the information that can be obtained from satellite images and, if required, from a comparison with basic technical features from known nuclear facilities. The challenge for the methodology is therefore to generate a maximum amount of information without any reliable information on the ground. Against the background of the detection of clandestine nuclear activities by individual countries within the framework of safeguards and the verification projects of the NPT, there is in such cases generally a lack of such reliable information.

Other criteria for selecting NFRPC Esfahan as a study area are the present safeguards requirements on the part of the IAEA, as well as the fact that in the past few years significant changes have been made to the facility. The following

S. Nussbaum, G. Menz, *Object-Based Image Analysis and Treaty Verification*,

© Springer Science+Business Media B.V. 2008

investigations are intended to indicate the extent to which certain objects at NFRPC can be classified and changes detected.

In this context, issues of automating the operating process, standardization, temporal and spatial transferability of classification models and quality assessments will be discussed.

7.1.1 Standardized Segmentation

The first step in object-based image analysis using eCognition software is multiscale segmentation, which groups individual pixels into objects. The mode of operation and the algorithm for multiscale segmentation have been derived and discussed in detail in Section 3.2. The following chapter will discuss the application of a standardized approach to the data available.

In the present study, the multiscale segmentation has been standardized in part. This was done by establishing a segmentation methodology which is applied to all data sources. The methodology, determined by preliminary tests, was applied once manually to the QuickBird-2 reference data set for NFRPC Esfahan from July 2002, recorded and then automatically transferred to other data sets. This method was developed by performing several test segmentations with different parameters.

The *scale parameter* will first be discussed in detail. This parameter indirectly determines the size of the individual objects, or put more simply, it permits a small scale parameter to extract small objects, and vice versa for large objects. In order to include objects of all sizes, a hierarchical network of various segmentation levels is set up. In the test runs with the reference data set it became apparent that no significantly different result was obtained in object extraction whether a scale parameter of 15, 25 or 35 was selected. Although in some cases objects were delineated better than others in the different scale levels, nevertheless parameter 25 achieved on average a good result. In this way a hierarchical network of objects was generated in the scale levels 25, 50, 75, . . ., 200, 225, 250. It is assumed that each real-world object in the picture of interest is extracted satisfactorily in one of these segmentation levels. For these spatially highly resolved data and in view of the aims of the present study it is not necessary to choose a parameter larger than 250. It would undoubtedly be possible to improve the segmentation by manual interventions, but this would not further the goal of automation. Segmentation should be standardized and should be transferable to other data sets.

In addition to the scale parameter, it became apparent that a weighting of the *colour* and *shape* parameters in a ratio of 0.7:0.3 is very well suited for the purposes of this study. In the case of the shape parameter, a weighting of *smoothness* to *compactness* in a ratio of 0.5:0.5 is a good measure.

Figure 7.1 shows the influence of the shape parameters of *smoothness* and *compactness* on the result of the segmentation. The higher weighting of the factor of compactness results in more compact segments, which is reflected, for example, in a

Segmentation Parameters

Scale	200
Color	0.7
Shape	0.3
Smoothness	0.9
Compactness	0.1

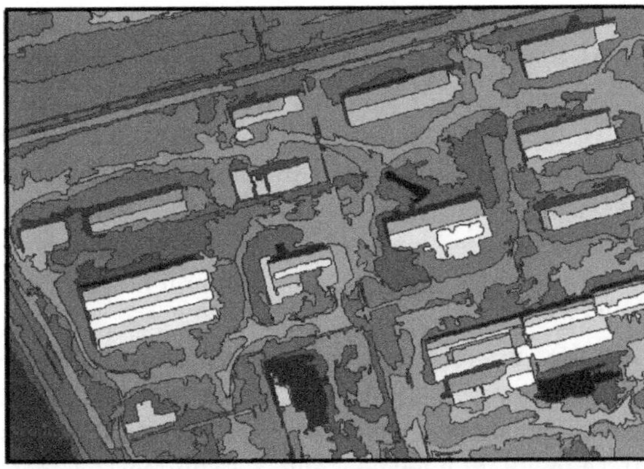

Original image panchromatically sharpened

Colour Composite	R: B1
	G: B2
	B: B3

Segmentation Parameters

Scale	200
Color	0.7
Shape	0.3
Smoothness	0.5
Compactness	0.5

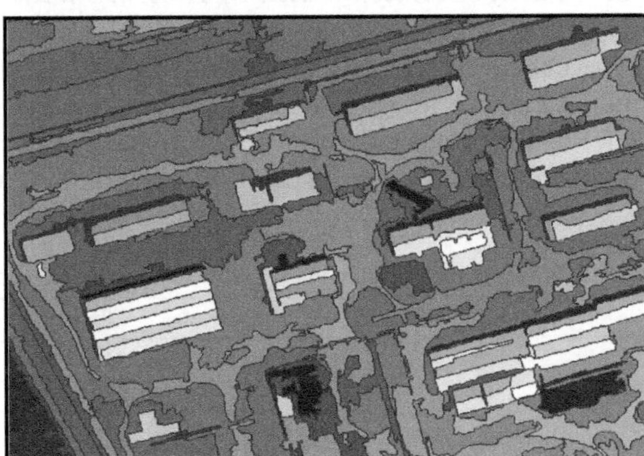

Fig. 7.1 Comparison of the segmentation results with altered shape parameters. The red arrows in the reference image show clear differences as an example. Original Data/Product © Quickbird/Digital Globe; 2002; Distributed by Eurimage

clear definition of the buildings. Even smaller buildings are retained in a higher scale level. However, it can also be seen that in spite of the shift of the shape parameters some segments display hardly any differences. This confirms the impression that small to moderate changes in the parameters do not have any significant influence on the result. The situation is different with the principal parameters of *colour* and *shape*, which yield strikingly different results depending on the setting. All the observations reported here refer to the reference data set used. It cannot be ruled out that different data material with different heterogeneity or spatial resolution would provide greatly diverging results.

For the further procedure in the present study, which applies the approach of automated image analysis, this means that all data sets are segmented in the same way with the above-mentioned parameters to form a hierarchical network consisting of 10 segmentation levels. Table 7.1 summarizes these parameters once again.

With the chosen method of standardized multiscale segmentation it cannot be ruled out that a better definition of the individual objects could be achieved by different or varying parameters. Nevertheless, this was not implemented since a standardization of the parameters created a common basis for all data sets and their further processing. This permits a first objective, comparative estimation of quality for the further processing steps and the results obtained. A conscious decision was thus made not to implement the optimum for the individual steps in the segmentation applied. Against the background of systemization, automation, transferability, representation of all significant objects, subsequent classification and quality assessment, this type of segmentation can, however, be regarded as a very good alternative.

7.1.2 Feature Analysis and Modelling with SEaTH

The theoretical background to semantic modelling has already been mentioned in Section 2.3 and its implementation in eCognition software discussed in Section 3.3. The following chapter describes the creation of a network of semantic features for the QuickBird-2 scene of NFRPC Esfahan in 2002. The starting point for the

Table 7.1 Systematized parameters for multiscale segmentation

	Scale Parameter	Color	Shape	Smoothness	Compactness
Level 1	25	0.7	0.3	0.5	0.5
Level 2	50	0.7	0.3	0.5	0.5
Level 3	75	0.7	0.3	0.5	0.5
Level 4	100	0.7	0.3	0.5	0.5
Level 5	125	0.7	0.3	0.5	0.5
Level 6	150	0.7	0.3	0.5	0.5
Level 7	175	0.7	0.3	0.5	0.5
Level 8	200	0.7	0.3	0.5	0.5
Level 9	225	0.7	0.3	0.5	0.5
Level 10	250	0.7	0.3	0.5	0.5

following study is a subset (3000 × 3000 pixels) of the original data set showing the Esfahan nuclear research centre. This data set has been pan sharpened by the *à trous cubic spline* filter method (cf. Section 6.2) and represents the reference data set for the following work (cf. Section 7.2). The figure shows a true colour combination of channels 3, 2, 1.

Fig. 7.2 QuickBird scene of NFRPC Esfahan, July 2002. Original Data/Product © Quickbird/Digital Globe; 2002; Distributed by Eurimage

7.1.2.1 Modelling the NFRPC Esfahan (July 2002)

In the following studies all features and thresholds for the semantic classification where determined using SEaTH (cf. Chapter 5). Table 7.2 gives an overview of all the features used for semantic modelling. A description of these features can be found in Section 3.3. In general, spectral features, shape features, textures, a vegetation index (NDVI) and ratios will be used for this purpose. A conscious decision was taken not to include features of specific satellite scenes, such as absolute position data for individual objects. These features are usually only valid for the image being studied and therefore are difficult to transfer to other satellite data. First of all, the fourth scale level was modelled for the satellite image of NFRPC of 2002. In this image, the buildings, vegetation, and also roads are well represented as image objects by the segmentation. In the first step, training objects (*samples*) of the individual classes are defined in eCognition for the SEaTH analysis program. About 3% of the total objects are identified as such training objects. Care should be taken that representative objects are selected. Furthermore, the number of samples of the individual classes should not differ greatly since a different number of samples for individual object classes that are difficult to differentiate would have a significant influence on the result of the SEaTH analysis. However, this does not make any significant difference for classes which can in principle be readily separated. If necessary, and if there is a known a priori probability of individual classes, the number of samples can be modified according to this a priori probability. Since the user of a classification does not as a rule know the separability and the a priori probabilities, the same number of samples per class should consequently be analysed with SEaTH.

After the samples have been selected in eCognition, a feature data set containing all 70 features (cf. Table 7.2) is extracted for them and then analysed with the SEaTH program. The results of the SEaTH analysis are converted into a classification model in eCognition. This model is then examined in later studies for its spatial and temporal transferability with respect to other satellite scenes.

The tables in Annex A show a section of the result of the SEaTH analysis for the fourth scale level of NFRPC Esfahan, July 2002. SEaTH makes it possible to analyse a large quantity of features and object classes[1] in a very short time. In this way, it is very easy to derive the ideal features with the associated thresholds for an object-based classification of the scene under consideration. The tables show the separability of the classes of **buildings** and **vegetation** (Table A.1), **buildings** and **roads** (Table A.1), **buildings** and **background** (Table A.1), **vegetation** and **roads** (Table A.1), **roads** and **background** (Table A.2) as well as **vegetation** and **background** (Table A.2). An interpretation of the tables now permits a statistically optimized classification model to be set up in eCognition. All 70 features for all objects are analysed with SEaTH (cf. Table 7.2). For reasons of space, the tables only show the "best" features according to the separability criterion. The separability has a dynamic range of [0.2] and, as already discussed in Chapter 5, SEaTH always

[1] The number of features and object classes can be extended as required.

Table 7.2 Features used for the semantic modelling of Esfahan, July 2002

ID	Feature	ID	Feature
1	Mean B1	36	Ratio B4
2	Stddev B1	37	GLCM Homogeneity (all dir.) B4
3	Ratio B1	38	GLCM Contrast (all dir.) B4
4	GLCM Homogeneity (all dir.) B1	39	GLCM Dissimilarity (all dir.) B4
5	GLCM Contrast (all dir.) B1	40	GLCM Entropy (all dir.) B4
6	GLCM Dissimilarity (all dir.) B1	41	GLCM Ang. 2nd moment (all dir.) B4
7	GLCM Entropy (all dir.) B1	42	GLCM Mean (all dir.) B4
8	GLCM Ang. 2nd moment (all dir.) B1	43	GLCM StdDev (all dir.) B4
9	GLCM Mean (all dir.) B1	44	GLCM Correlation (all dir.) B4
10	GLCM StdDev (all dir.) B1	45	Mean NDVI
11	GLCM Correlation (all dir.) B1	46	Stddev NDVI
12	Mean B2	47	GLCM Homogeneity (all dir.) NDVI
13	Stddev B2	48	GLCM Contrast (all dir.) NDVI
14	Ratio B2	49	GLCM Dissimilarity (all dir.) NDVI
15	GLCM Homogeneity (all dir.) B2	50	GLCM Entropy (all dir.) NDVI
16	GLCM Contrast (all dir.) B2	51	GLCM Ang. 2nd moment (all dir.) NDVI
17	GLCM Dissimilarity (all dir.) B2	52	GLCM Mean (all dir.) NDVI
18	GLCM Entropy (all dir.) B2	53	GLCM StdDev (all dir.) NDVI
19	GLCM Ang. 2nd moment (all dir.) B2	54	GLCM Correlation (all dir.) NDVI
20	GLCM Mean (all dir.) B2	55	Area (Pxl)
21	GLCM StdDev (all dir.) B2	56	Length (Pxl)
22	GLCM Correlation (all dir.) B2	57	Width (Pxl)
23	Mean B3	58	Length/width
24	Stddev B3	59	Compactness
25	Ratio B3	60	Elliptic Fit
26	GLCM Homogeneity (all dir.) B3	61	Rectangular Fit
27	GLCM Contrast (all dir.) B3	62	Border length (Pxl)
28	GLCM Dissimilarity (all dir.) B3	63	Shape index
29	GLCM Entropy (all dir.) B3	64	Density
30	GLCM Ang. 2nd moment (all dir.) B3	65	Main direction
31	GLCM Mean (all dir.) B3	66	Asymmetry
32	GLCM StdDev (all dir.) B3	67	Rel. Border to Brighter Neighbors B1
33	GLCM Correlation (all dir.) B3	68	Rel. Border to Brighter Neighbors B2
34	Mean B4	69	Rel. Border to Brighter Neighbors B3
35	Stddev B4	70	Rel. Border to Brighter Neighbors B4

mathematically provides two solutions for the determination of the thresholds $S1$ and $S2$. The final column show the threshold value, x, which is between the two means of the probability distribution. The sign makes it easier to establish the membership functions in eCognition. Table A.1 can be taken as an example. The first row can be interpreted as follows. The **buildings** class can be best separated from the **vegetation** class by the feature **mean B1** (separability = 1.74). All objects with feature characteristics of **mean B1** greater (sign = great) than 428.6 are assigned to **the buildings** class. A comparative interpretation of all tables finally permits a semantic model to be set up. In doing so, it is not always the best features that are used but also those that permit good separability for several classes. In this way, it is possible to reduce the number of features used for semantic modelling since this potentially improves the transferability of the model.

The class names are self-explanatory, with the exception of the **background** class. The objects in the background are all the objects that do not belong to any of the other classes. In order to distinguish the other classes such as **buildings** or **vegetation** from the background, it is, however, necessary to know which features permit this separation.

There is no complete separability of individual classes (separability = 2) for any of the defined classes. There are, however, values of up to 1.85, which permit very good separation since there is only a slight overlap in the probability distribution of the classes under consideration. The possibility of a misclassification is minimized by the automatic determination of thresholds. A separation between two classes in the feature investigated is not statistically possible to a better extent that by the threshold value determined by SEaTH as a function of the samples used.

It can furthermore be seen that the classes of **buildings**, **vegetation** and **background** can mostly be well separated by spectral features. The situation is different for the **roads** class. A separation of this class from the classes of **buildings** and **background** cannot be adequately performed with spectral features. It is clearly shape features that enable an optimal separation to be made.

Table 7.3 shows that the classification rules resulting from the SEaTH analysis for objects of the classes of **buildings**, **vegetation**, **roads** and **background** for the fourth scale level. Against the background of temporal and spatial transferability of the classification model a small number of features is desirable for describing the classes.

After the semantic modelling of the object classes in the fourth scale level, smaller objects can now be classified and modelled on lower levels. The deep shadows of the buildings can be readily recognized as objects in scale level two by the standardized segmentation. Table A.2 shows the results of the SEaTH analysis for the **shadows** class. In the shadows class, a separation of this class from the classes of **buildings** and **background** is necessary since in the higher level four the shadows are still classified as objects of the **buildings** and **background** classes.

The evaluation of the results leads to the establishment of classification rules for scale level two (cf. Table 7.4). Here the **shadows** class is differentiated from the

Table 7.3 Classification rules of the classes of buildings, vegetation, roads and background for the fourth scale level of the scene NFRPC Esfahan, July 2002

Class	Feature	Threshold
Buildings Lvl 4	Ratio B2	> 0.306
	Mean NDVI	< 0.021
	Ratio B4	< 0.262
	Length	< 172
Vegetation Lvl 4	Mean B1	< 394
	Mean NDVI	> 0.056
Roads Lvl 4	Density	< 1.3
	Length	> 172
	Mean NDVI	< 0.03
Background Lvl 4	*not*	Buildings
	not	Vegetation
	not	Roads

other classes by the relation of its objects to the neighbouring objects as well as by the spectral feature *ratio B1*. Due to the hierarchical structure of the individual scale levels, it is possible to take over object classes already classified in the fourth scale level.

In the first segmentation level, the smallest objects can be described and classified. In the area of the Esfahan Nuclear Research Centre under consideration there are fences, walls and other linear boundary structures as well as a branched pipe system. Studies with SEaTH have shown that these similar structures cannot be satisfactorily separated from each other,[2] but that they can be distinguished from their surroundings. This fact leads to a rather unusual class name **walls, fences and pipe systems**. Table A.3 shows the significant results of the SEaTH analysis. The texture feature *GLCM correlation (all dir.) B3* and also *NDVI* separate the **walls, fences and pipe systems** very well from the **roads**, **vegetation** and **background**. In accordance with these findings, classification rules were set up for the first scale level (cf. Table 7.5).

Table 7.4 Classification rules of the classes buildings, vegetation, roads and background for the second scale level of the scene NFRPC Esfahan, July 2002

Class	Feature	Threshold
Shadow Lvl2	Ratio B1	> 0.199
	Relative border to brighter neighbors	1
Buildings Lvl2	Existence of Buildings Lvl4 super-objects (2)	1
	not Shadows Lvl2	1
Vegetation Lvl2	Existence of Vegetation Lvl4 super-objects (2)	1
Roads Lvl2	Existence of Roads Lvl4 super-objects (2)	1
Background	*not*	Buildings Lvl2
	not	Shadows Lvl2
	not	Vegetation Lvl2
	not	Roads Lvl2

[2] The spatial resolution of the dataset is to low.

Table 7.5 Classification rules of the classes of walls, fences and pipe systems, shadows, buildings, vegetation, roads and background for the first scale level of the scene NFRPC Esfahan, July 2002

Class	Feature	Threshold
Walls, Fences and Pipe Systems	GLCM Correlation (all dir.), B3	< 0.57
	Mean NDVI	< 0.036
	Stddev B1	> 13.4
Shadows Lvl1	Existence of Shadows Lvl2 super-objects (1)	1
Buildings Lvl1	Existence of Buildings Lvl2 super-objects (1)	1
Vegetation Lvl1	Existence of Vegetation Lvl2 super-objects (1)	1
Roads Lvl1	Existence of Roads Lvl2 super-objects (1)	1
	not	Walls, Fences and Pipe systems
Background Lvl1	*not*	Walls, Fences and Pipe systems
	not	Buildings Lvl2
	not	Shadows Lvl2
	not	Vegetation Lvl2
	not	Roads Lvl2

Figure 7.3 shows the entire classification model with the classification rules determined by SEaTH for the RS data of the NFRPC Esfahan, July 2002. This model will be used in the following chapter to classify the area under investigation and assess the quality of this classification.

7.1.3 Classification – Results and Quality Assessment

The semantic classification of the satellite scene is performed hierarchically for each scale level modelled. In this way, first the fourth, then the second and finally the first scale level is classified. Due to the network nature of the model, the information from the upper levels is known in the lower levels. This enables multiscale objects to be classified in one operation.

The semantic model determined by SEaTH (cf. Chapter 5) is used to describe the individual classes (cf. Fig. 7.3). The classification is performed according to a fixed rule base so that the influence of fuzzy logic is deliberately ruled out. The thresholds of the individual features used for describing the classes are defined as clear boundaries without any transition region. The object either belongs to this class (membership = 1) or it does not (membership = 0). The deliberate decision not to use fuzzy logic is based on the fact that the thresholds found by SEaTH are statistically optimal. Any shift or expansion leads to an increase in misclassifications. Accordingly, all the defined membership functions follow the pattern shown in Fig. 7.4.

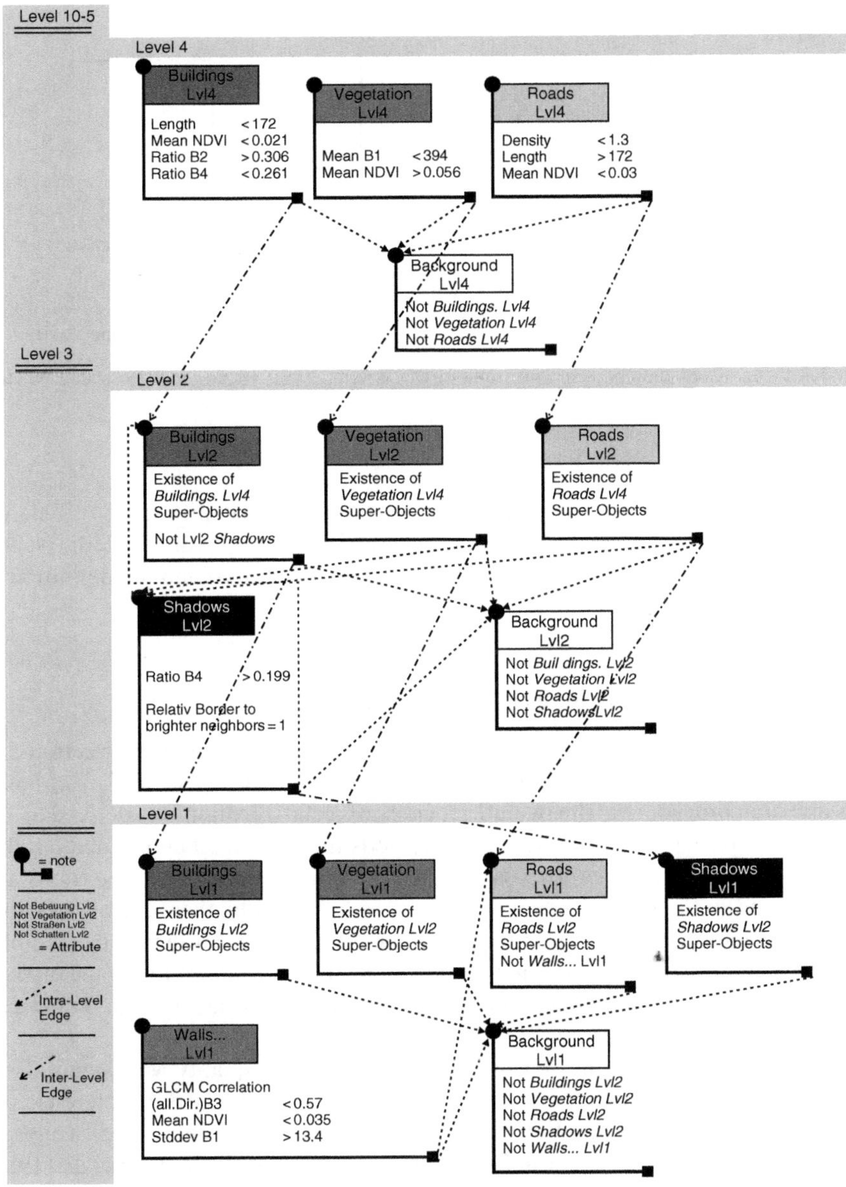

Fig. 7.3 Semantic feature model of the satellite scene NFRPC Esfahan, July 2002. The object classes and classification rules are shown in the individual scale levels used

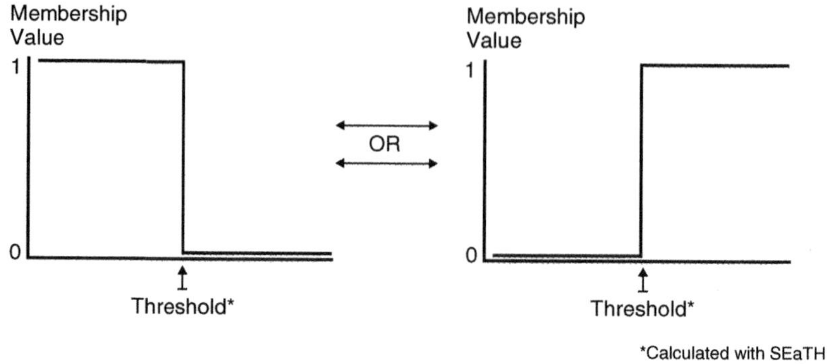

Fig. 7.4 Generalized membership functions of the semantic classification of the scene NFRPC Esfahan, July 2002

Figure 7.5 shows the result of the classification of the satellite scene of Esfahan from July 2002. Figure 7.6 shows an enlarged section of the classified satellite scene. In this section the individual classes can be easily recognized and the definition of the facility by walls is also clear.

7.1.3.1 Quality Assessment of the Classification

The quality assessment measures used in this thesis are discussed in Section 5.2. These measures were originally developed for the quality assessment of pixel-based classification procedures. The overall accuracy of a classification was derived on the basis of reference pixels (samples). The methods for the correct choice and number of these samples have been discussed in the literature on numerous occasions and can be found in the references given in Section 5.2. Finally, there are also various established methods of selecting these samples. In the present thesis, the samples are selected at random. A minimum number of samples is chosen for all the classes. Once the samples have been determined, the quality can be assessed in a next step by using the selected measures.

An important point is the statistical validity of such a quality assessment. How representative of the whole picture is a quality assessment based on reference objects? The answer to this question is closed related to the number of samples chosen. In their work, authors such as van Genderen *et al.* (1978) or Rosenfield *et al.* (1982) specify minimum numbers of samples for pixel-based methods in order to achieve a certain level of significance. A different approach is taken in the present thesis in order to determine the validity of the calculated quality measures. One reason for this is the new, object-based approach so that existing statements about pixel numbers cannot be applied. In principle, determination of the precision of the objects is performed in exactly the same way as for pixels with the sole, but major, difference that the number of objects in a scene is always significantly lower than the number of pixels. Consequently, the possible number of samples also changes.

Fig. 7.5 Result of the semantic classification of the satellite scene of NFRPC Esfahan, July 2002. Original Data/Product © Quickbird/Digital Globe; 2002; Distributed by Eurimage

The reference objects are visually determined since, as already mentioned, there is no reliable ground information for the area under investigation. The spatial resolution of the satellite images is, however, so high that reference objects for the selected classes can be reliably identified without difficulty.

Canty (1999) specifies the validity of a quality assessment via the standard deviation of the misclassification rate. It is thus possible to make statistically meaningful statements about the probability of incorrectly classified objects. This methodology – a derivation can be found in Canty (1999) – can also be used to determine the standard deviation of the correctly classified objects. This merely involves exchanging Canty's initial definitions of incorrectly and correctly classified. The estimated standard deviation finally gives the correctly classified objects as

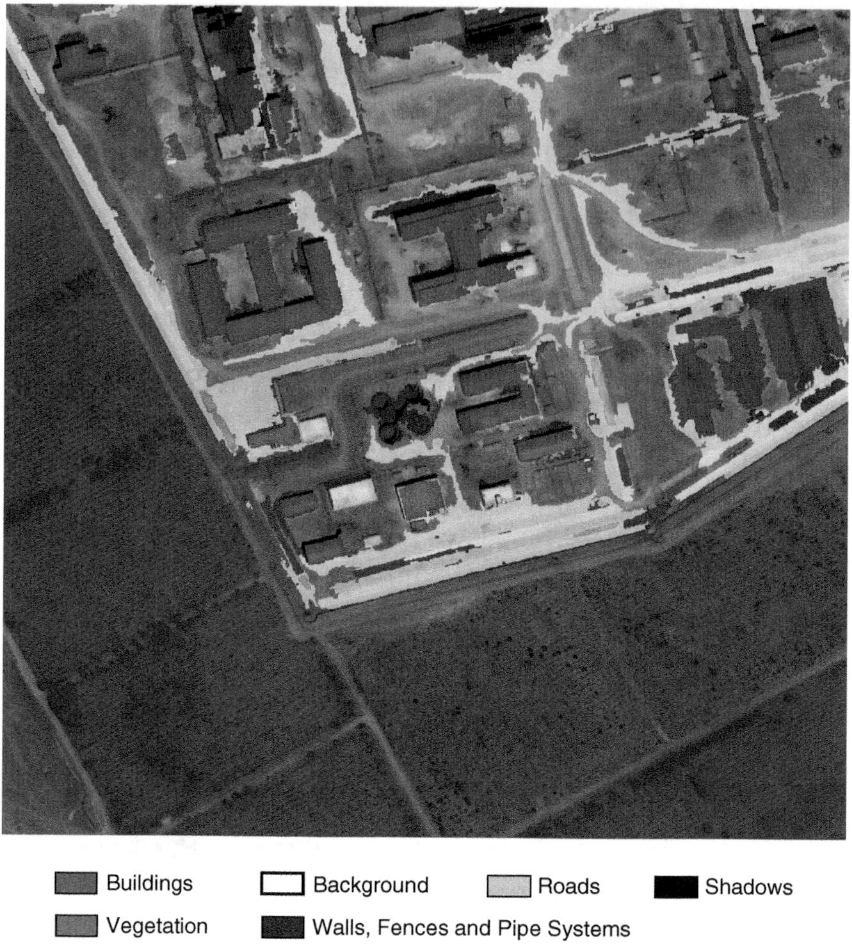

Fig. 7.6 Subset of the classification of the satellite scene of NFRPC Esfahan, July 2002. Original Data/Product © Quickbird/Digital Globe; 2002; Distributed by Eurimage

$$\sigma = \sqrt{\frac{y(n-y)}{n^3}}. \tag{7.1}$$

where n designates the total number of samples and y the correctly classified objects.

Table 7.6 shows the quality assessments of the object-based classification of NFRPC Esfahan, July 2002. The quality is determined on the lowest scale level since all object classes are represented there. This permits a holistic quality assessment. At the significance level of 95%, corresponding to 2σ, an *overall accuracy* between 89% and 93% is reached. For the **background** class the *producer accuracy* on the 95% significance level is 91–99%, for **buildings** 100%, for **roads** 69–89%,

Table 7.6 Quality assessment of the object-based classification of the satellite scene of NFRPC Esfahan, July 2002

Classified Objects	Reference Objects						
	Background	Buildings	Roads	Vegetation	Shadows	Walls...	$\sum$
Background	76	0	10	2	2	10	100
Buildings	0	78	6	0	0	0	84
Roads	2	0	60	0	6	0	68
Vegetation	0	0	0	100	0	0	100
Shadows	0	0	0	0	44	0	44
Walls...	2	0	0	0	0	40	42
$\sum$	80	78	76	102	52	50	**438**
Producer Accuracy	0.95	1.00	0.79	0.98	0.85	0.80	
User Accuracy	0.76	0.93	0.88	1.00	1.00	0.95	
Overall Accuracy	0.91						
Producer Accuracy σ	0.02	0	0.05	0.01	0.05	0.06	
Producer Accuracy 95%	0.91–0.99	1	0.69–0.89	0.96–1	0.75–0.95	0.68–0.92	
User Accuracy σ	0.04	0.03	0.04	1	1	0.03	
User Accuracy 95%	0.68–0.84	0.87–0.99	0.80–0.96	1	1	0.89–1	
Overall Accuracy σ	0.01						
Overall Accuracy 95%	0.89–0.93						

for **vegetation** 96–100%, for **shadows** between 75% and 95% and for **walls, fences and pipe systems** 68–92%. This means that the classification model correctly classifies the **background, buildings** and **vegetation** classes with 95% probability as at least 90% or better. The **roads, shadows** and also **walls, fences and pipe systems** classes display a wide variation range in σ of up to 6%.

The values for *user accuracy* can similarly be read off and interpreted from Table 7.6. By far the lowest user accuracy is provided by the **background** class with 68–84% on the 95% significance level. A total of 438 reference areas were used for quality determination, which corresponds to more than 2% of the total objects.

A pixel-based supervised classification was used to compare the quality of the results. The same classes are defined and samples are also selected. On the basis of these samples a supervised classification is performed with the *maximum likelihood* classifier. The results of this brief comparability test are only summarized here. On the 95% significance level the overall accuracy amounts to 79–81%. The **roads** class displays a *user and producer accuracy* of less than 30% and the **walls, fences and pipe systems** class only achieves a rather worrying 8.2% in user accuracy. This shows the superiority of a broader feature basis very clearly since, as already surmised, spectral features alone are not sufficient to identify certain object classes.

7.2 Temporal Transferability of a Reference Classification Model

Since typical features for individual object classes or facility types can be found with the selected methodology, interest is now focused on the possible transferability of these reference models to other RS data or other periods of time. If this transferability is found to be satisfactory, then information from RS data can be made available in a fast and simple manner.

In this chapter, the reference classification model developed for NFRPC Esfahan (July 2002) is transferred to satellite scenes of the same site at later points in time. The goal is therefore a verification of the temporal transferability of the classification model and the rule base with features and threshold values.

7.2.1 Site – NFRPC Esfahan, 2003

The first investigation period is a QuickBird-2 satellite scene of NFRPC Esfahan in July 2003. The temporal distance from the reference model is almost exactly 1 year. Figure 7.7 shows the initial image for the subsequent investigations after preprocessing in a true colour representation of channels 3, 2, 1. From a spatial point of view, the reference scene of 2002 and that of 2003 are congruent. After preprocessing, the four channels of the satellite scene and the NDVI are loaded in eCognition. The object extraction is performed by the standardized segmentation determined in Section 7.1.1. This has been recorded and can be transferred and applied without difficulty. After segmentation, the reference classification model can be imported and also applied without any problem. In the first step, all the settings of the model are taken over completely. This refers to parameters such as hierarchical structure, the features and the threshold values. The results of the classification of the scene from Esfahan July 2003 with the reference model are shown in Fig. 7.8.

The **vegetation**, **shadows** and a large proportion of the **buildings** can be readily identified and classified from the reference model. The same is true of large, asphalted **roads** and some **walls, fences and pipe systems**. However, there is a striking lack of small roads and paths as well as some boundary structures and buildings. On the whole, however, in view of the temporal distance of 1 year, the result can be regarded as surprisingly positive with respect to the complete transferability of the reference model. Nevertheless, the overall result is not yet satisfactory.

It can be conjectured that the threshold values of the features optimized for the year 2002 cannot be transferred to the 2003 satellite scene without being adapted. In the following, some of the attributes of the features will be slightly modified. Since some objects actually present were not recognized in all of the classes, it is apparent that the threshold values should be slightly increased. Since the NDVI is a frequently used feature for determining the classes and this feature is liable to change over time we decided to modify it slightly. It is apparent that the threshold value of the NDVI only needs to be adjusted upward slightly. Table 7.7 shows the changes in the NDVI threshold values in direct comparison with 2002. The values of

Fig. 7.7 QuickBird scene of NFRPC Esfahan, July 2003. Original Data/Product © Quickbird/Digital Globe; 2003; Distributed by Eurimage

the NDVI range from -1 to 1. A maximum change of 0.008, as made for the class of buildings and roads, can certainly be regarded as slight. The threshold values of the standard deviation feature in the class of walls, fences and pipe systems is reduced to 7. Otherwise, the reference model is taken over in its entirety.

Figure 7.9 now shows the satisfactory result in the object-based classification with slightly adapted thresholds for two of the features. In order to evaluate and to

Fig. 7.8 Result of the classification of the satellite scene of NFRPC Esfahan, July 2003, with the 2002 reference model. Original Data/Product © Quickbird/Digital Globe; 2003; Distributed by Eurimage

compare the classification results, the same procedure will be applied as described in detail in Section 7.1.3. Table 7.8 shows the result of this quality assessment.

The *overall accuracy* is about 84%–92% at the 95% significance level. A very good *producer accuracy* with a low sigma value is achieved by the classes of

Table 7.7 Fitting the NDVI threshold value in the classification model

	2002	2003
Buildings	0.021	0.028
Walls, fences and pipe systems	0.035	0.037
Roads	0.030	0.038

Fig. 7.9 Result of the semantic classification of the satellite scene of NFRPC Esfahan, July 2003, with the adjusted NDVI threshold value. Original Data/Product © Quickbird/Digital Globe; 2003; Distributed by Eurimage

background (100%), **buildings** (94%–100%) and **vegetation** (90%–98%). Smaller objects, therefore more difficult to identify, achieve a lower degree of quality and have a correspondingly higher sigma value. This includes the classes of **roads**, **shadows** and also **walls, fences and pipe systems**. **Vegetation** is the class with the highest quality in producer and user accuracy.

Since the semantic model for classification is the slightly modified 2002 reference model, the quality values achieved are very satisfactory. Temporal transferability is therefore possible in order to obtain initial information from RS data very rapidly. The small modifications to the threshold of the NDVI only take a few minutes.

Table 7.8 Quality assessment of the object-based classification of the satellite scene of NFRPC Esfahan, July 2003

Classified Objects	Reference Objects						
	Background	Buildings	Roads	Vegetation	Shadows	Walls...	$\sum$
Background	84	0	14	6	0	8	112
Buildings	0	84	2	0	6	0	92
Roads	0	0	54	0	0	6	60
Vegetation	0	0	0	98	0	0	98
Shadows	0	2	2	0	30	0	34
Walls...	0	0	2	0	4	46	52
$\sum$	84	86	74	104	40	60	448
Producer Accuracy	1.00	0.98	0.73	0.94	0.75	0.77	
User Accuracy	0.75	0.91	0.90	1.00	0.88	0.88	
Overall Accuracy	0.88						
Producer Accuracy σ	0	0.02	0.05	0.02	0.07	0.05	
Producer Accuracy 95%	1	0.94–1	0.63–0.83	0.90–0.98	0.61–0.92	0.67–0.87	
User Accuracy σ	0.04	0.03	0.04	0	0.06	0.04	
User Accuracy 95%	0.67–0.83	0.85–0.97	0.82–0.98	1	0.76–1	0.80–0.96	
Overall Accuracy σ	0.02						
Overall Accuracy 95%	0.84–0.92						

7.2.2 Site – NFRPC Esfahan, 2004

After July 2003, the temporal transferability of the reference model to a QuickBird-2 satellite scene of the NFRPC from May 2004 was additionally tested. The methodological procedure is the same as in Section 7.2.1 and will not be discussed in detail here. Figure 7.10 shows the preprocessed satellite image in a true colour representation.

Since the temporal distance from the reference model amounts to 2 years, as in the case of the 2003 scene, it is necessary to fit the threshold values. The features as such are retained. Table 7.9 gives a list of the adaptations of the individual threshold values. It is immediately obvious that only spectral features have to be changed, whereas shape and texture features appear to be transferable over a long period of time. The reason for the modification of several threshold values can also be the fact the scene was recorded in May and not in July like the other two. As in the case of 2003, it is mainly the NDVI that requires slight adjustment.

Figure 7.11 shows the result of the classification with the fitted model. The first impression is that of a good classification result even if it is apparent that there are more misclassifications in all object classes than for 2002 and 2003.

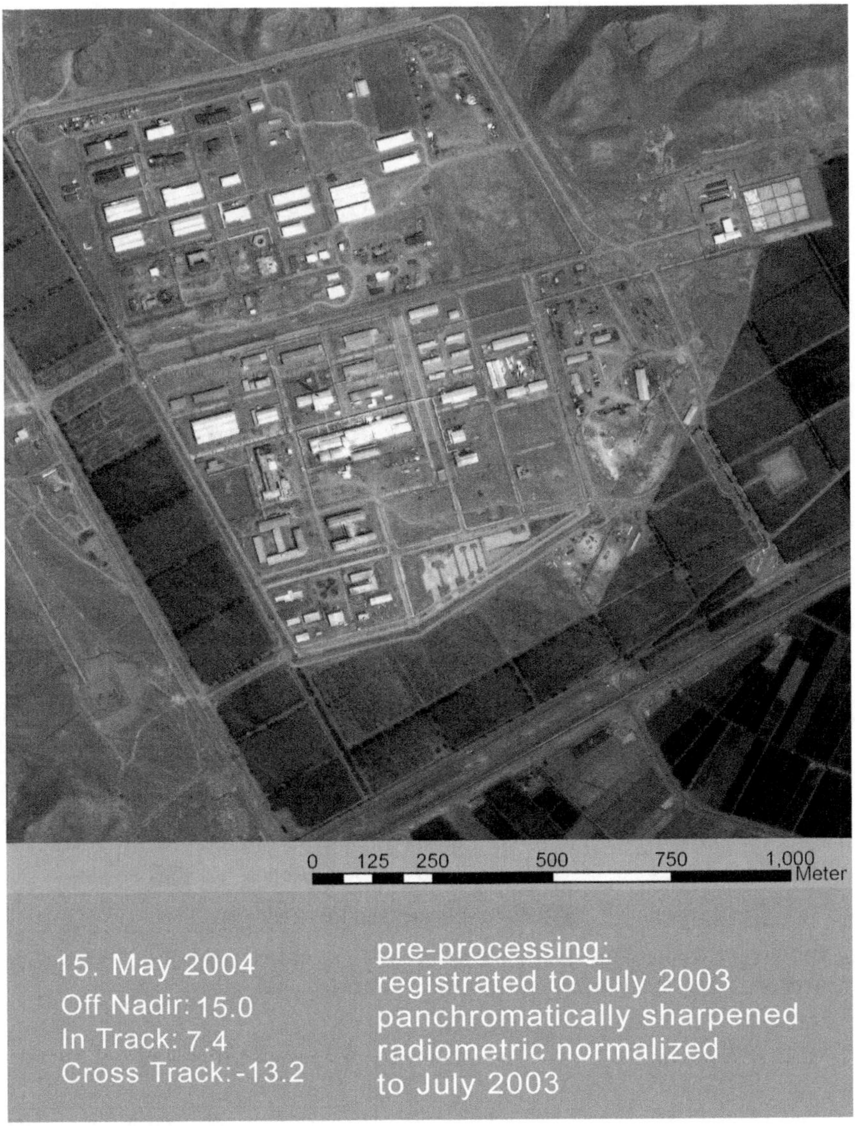

Fig. 7.10 QuickBird scene of NFRPC Esfahan, May 2004. Original Data/Product © Quickbird/Digital Globe; 2004; Distributed by Eurimage

Table 7.10 gives a quantitative assessment of the classification. The *overall accuracy* achieved of 83%–91% at the 95% significance level is certainly acceptable, although poorer than that achieved for the year 2003. The trend is also apparent for many other classes. The *user accuracy* of the classes **roads, vegetation, shadows** and **walls, fences and pipe systems** is surprisingly good. This leads to the conclusion that a very reliable classification has been made, i.e. objects are really only

Fig. 7.11 Result of the classification of the satellite scene of NFRPC Esfahan, May 2004. Original Data/Product © Quickbird/Digital Globe; 2003; Distributed by Eurimage

Table 7.9 Fitting the threshold values of the reference model to 2004

	2002	2004
Vegetation		
Mean Band 1	< 394	< 450
Roads		
Mean NDVI	< 0.03	< 0.04
Buildings		
Mean NDVI	< 0.021	< 0.04
Ratio Band 4	< 0.26	< 0.28
Walls...		
Mean NDVI	< 0.35	< 0.37
Stddev Band1	> 13.4	> 7

Table 7.10 Quality assessment of the object-based classification of the satellite scene of NFRPC Esfahan, May 2004

Classified Objects	Reference Objects						
	Background	Buildings	Roads	Vegetation	Shadows	Walls...	Σ
Background	78	8	14	4	0	8	112
Buildings	0	94	4	0	14	4	116
Roads	0	0	42	0	0	0	42
Vegetation	0	0	0	62	0	0	62
Shadows	0	0	0	0	48	0	48
Walls...	0	0	0	0		38	38
Σ	78	102	60	66	62	50	418
Producer Accuracy	1.00	0.92	0.70	0.94	0.77	0.76	
User Accuracy	0.70	0.81	1.00	1.00	1.00	1.00	
Overall Accuracy	0.87						
Producer Accuracy σ	0	0.03	0.06	0.03	0.05	0.06	
Producer Accuracy 95%	1	0.86–0.98	0.58–0.82	0.88–1.00	0.67–0.87	0.64–0.86	
User Accuracy σ	0.04	0.04	0	0	0	0	
User Accuracy 95%	0.62–0.78	0.73–0.89	1	1	1	1	
Overall Accuracy σ	0.02						
Overall Accuracy 95%	0.83–0.91						

assigned to the classes to which they actually belong. This brief example of the temporal transferability of the reference model concludes the detailed studies in this field. It is apparent that transferability can be achieved over a period of 2 years, even if the threshold values have to be adjusted. However, the quality of the classification decreases with increasing temporal distance. In the further course of this study, the spatial transferability of the reference model will be analysed on the basis of various regions.

7.3 Spatial Transferability of the Reference Classification Model

In addition to *temporal transferability* of the classification model, there may also be the possibility of *spatial transferability*. In the following chapter, it will be examined whether the reference model of NFRPC Esfahan from 2002 can be transferred to other sites. Subsets of the original QuickBird-2 full satellite image from 2002 were used as a first starting point. These images are in the close vicinity of NFRPC Esfahan, but outside the section of the image used in Section 7.1.

The approach is the same as that described in the previous chapters. The substes of the image were divided into objects by a standardized multiscale segmentation. The reference model was then applied to the individual scenes. Figure 7.12 shows the test regions used and the result of the classification. It becomes apparent that the **vegetation** and large **roads** and **buildings** are readily identified. In the central section, however, the small **buildings** of the settlement located there are not recognized at all. The classes of **shadows** and **walls, fences and pipe systems** also display a certain amount of misclassification.

As in the case of temporal transferability, it is obvious that the features are themselves well suited for describing the classes, but that the thresholds in general, and in particular those of the spectral features, require slight adaptation. This has been confirmed by further tests, which will not be described in detail here since different test areas are being considered. In this way, slight adjustment of the thresholds of the NDVI and the ratios leads to considerably improved results for these test areas.

In the next step, the reference model of NFRPC Esfahan was transferred to the region around Bushehr, Iran. It was still possible to ensure partial transferability since Bushehr is located in the same country and thus in the same (macro-)climate zone. In addition to the unchanged 2002 reference model, the class of **water** was

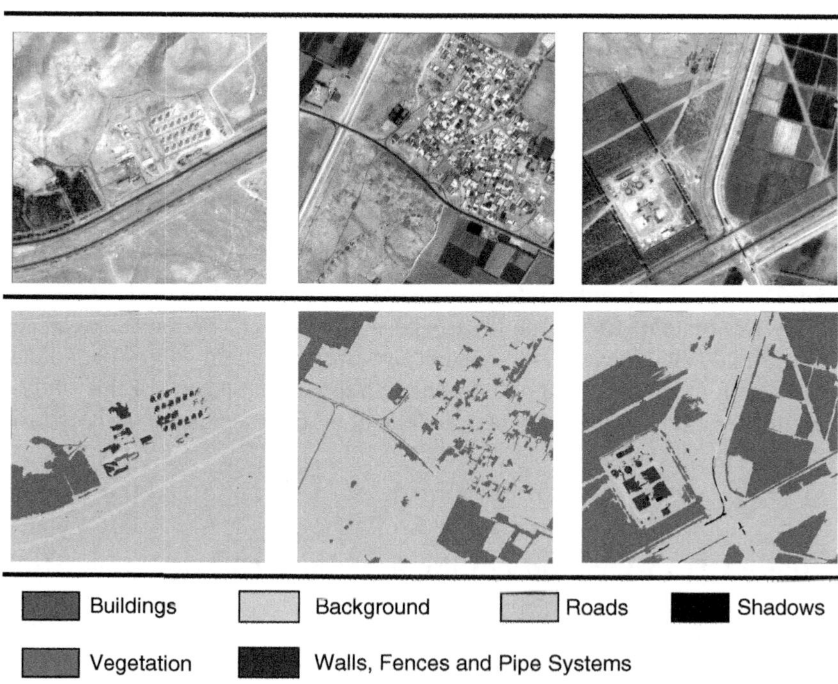

Fig. 7.12 Subsets of the image and corresponding classification results for test areas outside NFRPC Esfahan. Original Data/Product © Quickbird/Digital Globe; 2003; Distributed by Eurimage

modelled via the feature *mean in band 3*. This feature is known and permits a good separation of water from all other classes. The result of the classification (cf. Fig. 7.13) was unexpectedly good although the test area is a relatively long way from Esfahan. The large **buildings** (with the reactor) and also a complex **road** network can be identified here. The settlement with small buildings at the western edge of the scene was not recognized.

A test area in a very different spatial setting will now be considered. This is a nuclear power plant near Pickering in Canada. Apart from the spatial discrepancy, the data of this region were recorded by a different satellite, IKONOS. This satellite has a different spatial and spectral resolution from that of QuickBird-2 so that due to this fact the chance of good transferability was regarded as low. In addition to the reference model, a class of **water** was also modelled here via the *mean in band 3*.

The upper left image segment of Fig. 7.14 (A) shows the original image and the upper right segment (B) shows the result of the first classification. In spite of the enormous differences in spatial area and a different sensor the classes of **vegetation** and also some **roads** and **shadows** are correctly classified. **Buildings**, in contrast, are not recognized.

In the next step, it will be shown whether by adjusting some of the thresholds it is possible to achieve a better classification result for the **buildings** class. To this end, the threshold of the feature *mean NDVI* will be increased from 0.021 to 0.04 and that in *B2 ratio* reduced from 0.3 to 0.25. Figure 7.14 (C) shows the result of the subsequent classification. A large proportion of the buildings present are not correctly classified. However, in contrast to NFRPC Esfahan, no individual buildings can be separated on the site of the nuclear facility. One reason for this is that the entire area of the nuclear power plant is concreted and therefore is spectrally very similar to the

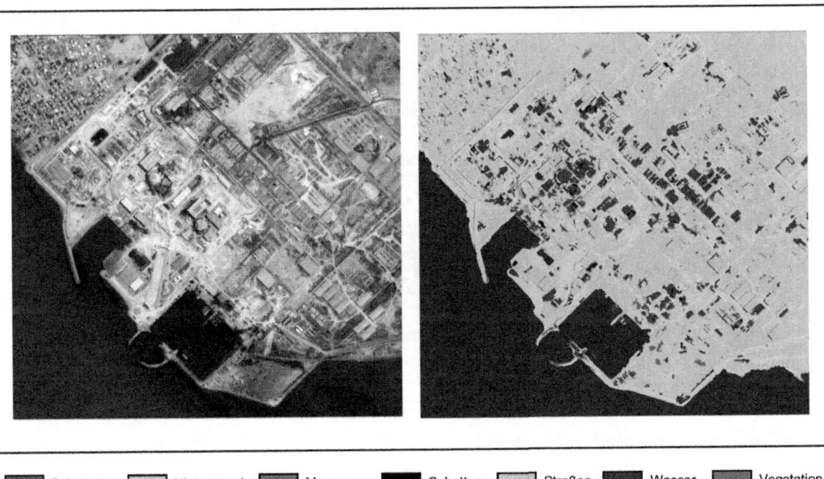

Fig. 7.13 Image subset and corresponding classification result for the Bushehr test area. Original Data/Product © Quickbird/Digital Globe; 2003; Distributed by Eurimage

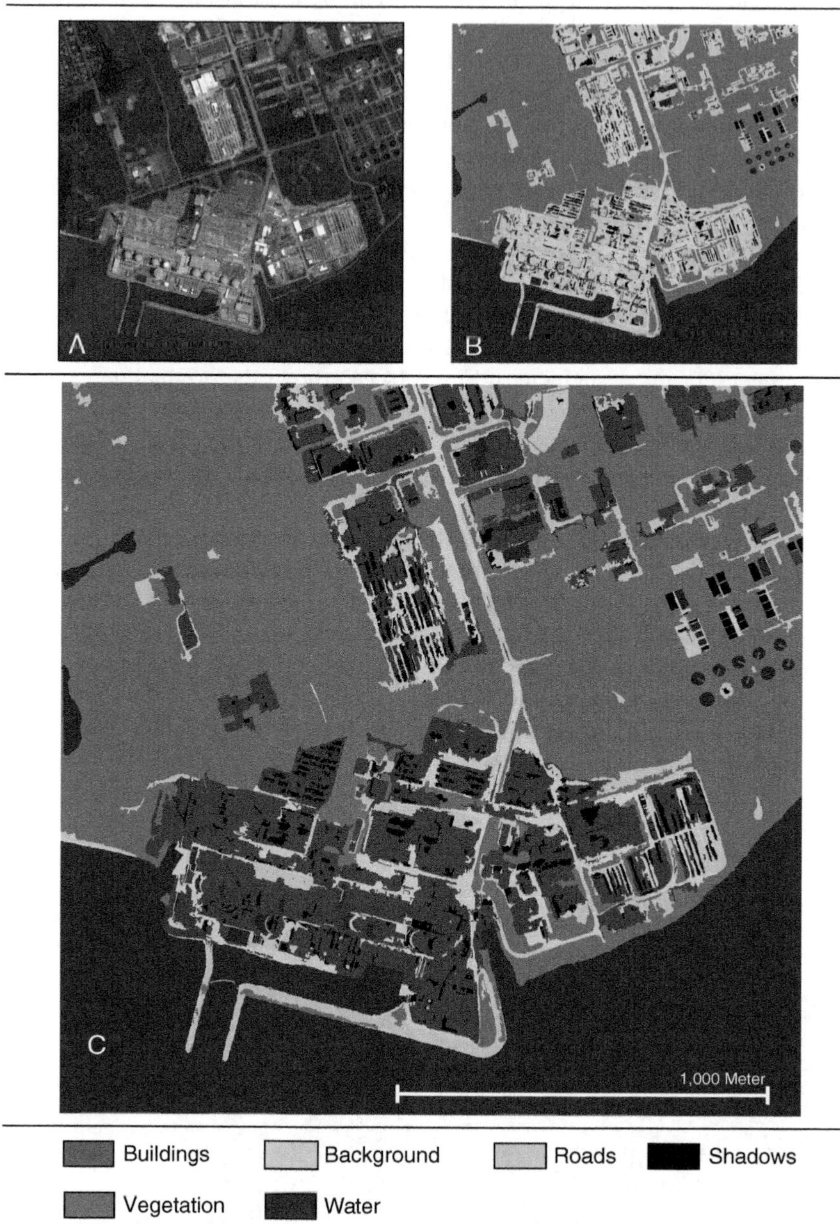

Fig. 7.14 Image subset and corresponding classification result for the Pickering test area in Canada. Includes material © GeoEye all rights reserved

buildings. In order to identify individual buildings, other features such as shapes or textures should be considered in the model.

As a result of this transferability test, it can nevertheless be confirmed that the model has a surprisingly good spatial transferability. In this case as well, it can be seen that the features found by SEaTH can be regarded as having more or less general validity in separating the classes identified. For transferability over a wide area, the threshold values of the individual classes and features require adjustment, which can, however, be achieved with little extra work and in a very short time. It is, however, understandable that complete transferability including the thresholds cannot be ensured. This corresponds to our expectations since some of the features, such as those of the NDVI or the band ratios, are dependent on factors such as the satellite sensor, climate zone or season. Nevertheless, the transfer was satisfactory as a whole and by means of a minor adjustment of the threshold values it was possible to obtain initial information from the satellite image in a rapid and simple manner. However, spatial transferability does not seem to be a suitable means of obtaining good or very good quality and information content.

7.4 Sites – Bandar Abbas, Bushehr and Natanz

The spatially very high-resolution QuickBird data that are mainly used in the present study are well suited for identifying individual objects in nuclear facilities. Data of a lower spatial resolution can be used to detect the facilities themselves. The latter have the advantage of being able to cover a much larger area in a single satellite scene and they can be obtained more cheaply. If the aim is monitoring in the sense of nuclear safeguards then the first objective is to detect the facilities themselves and then to analyse the findings in more detail using high-resolution data.

The aspect of identifying nuclear facilities with the example of Germany has already been examined in Niemeyer (2001). In the following, some facilities in Iran, represented by Aster satellite data, will be processed in an object-based manner using the SEaTH methodology. It will be analysed whether a reference model can be set up for detecting industrial facilities in wide-area ASTER satellite data. As an example, it will be shown whether the SEaTH methodology is also appropriate for spatially low-resolution data and how good the quality of the resulting model is. This chapter represents a slight digression from spatially high-resolution data, but nevertheless shows how the methodology can also be applied to other satellite data of different resolution. Furthermore, as already mentioned above, this would be the first step towards large-scale monitoring of a whole country.

The ASTER (Advanced Spaceborne Thermal Emission and Reflection Radiometer) satellite was launched in December 1999. The satellite has a broad spectral resolution ranging from visible light up to the thermal infrared region, divided into 14 bands. The spatial resolutions vary with the wavelength: 15 m in the visible and very near infrared (VNIR), 30 m in the short-wave infrared (SWIR) and 90 in the thermal infrared. The satellite-specific characteristics will not be discussed in detail here and readers are referred to the existing literature and to the relevant handbook, the *Aster User* Guide.[3]

The spatially most highly resolved data in the VNIR region and the SWIR data were used for the study. In a preprocessing step, the SWIR data were pan sharpened by the cubic spline filter method (Section 6.2) and then combined into a data set by so-called layer stacking with the VNIR data. The Iranian facilities of Bandar Abbas, Bushehr and Natanz were selected as the study areas. A brief description of the facilities can be found in Section 1.2. The facilities themselves only represent a small section of the satellite scene studied. The aim is to precisely identify this small region of the image with the aid of the object-based classification model determined by SEaTH. Figure 7.15 shows the original ASTER images after preprocessing. The figure presents a colour combination of the first three bands.

The procedure in this investigation is similar to that in the previous chapter where the NFRPC Esfahan was analysed and classified. In the first step, all three satellite images were divided into objects by a standardized segmentation. Attention was paid that the facility, or large parts of the facility, were represented in one segment. The result of the multiscale segmentation and also the segmentation parameters used are shown in Fig. 7.16. The segmentation parameters correspond to the standardized parameters which were also used for the analysis of NFRPC Esfahan (cf. Section 7.1.1). However, due to the lower spatial resolution, slightly different individual scale levels were selected. It was found that the facilities are well represented as objects on a scale factor of 75–100.

Modelling is performed after segmentation. The six classes of **soil**, **mountains**, **industrial sites**, **settlement**, **vegetation** and **water** are defined as well as training objects of the individual classes for feature analysis with SEaTH in all four images. The features of the training objects are combined to form a feature data set. A good cross section of all the data is thus obtained. The available feature basis is very broad due to the 9 bands used for the analysis of the ASTER data.

A *digital elevation model* (DEM) is included in the feature analysis as additional information. On the one hand, this is used to calculated features of elevation as well as additional features of slope gradient. The DEM is easy to generate since the ASTER satellite additionally has a backward scanner. The procedure for extracting DEMs from ASTER satellite images will not be described in detail here and the reader is referred to the ASTER User Guide.[4] The ASTER DTM tool was then used to calculate a digital terrain model (DTM) for the three satellite scenes. A total of 191 features (cf. Tables B.1 and B.2)are included in the analysis.

The results of the SEaTH analysis are compiled in Tables B.3–B.5. Only the "best" features in the separability criterion are shown. The first interpretation of the tables leads to the conclusion that most classes can be readily separated. The class of water achieves complete separability from all other classes (separability = 2). It is also noticeable that it is mainly texture features that permit the best separation.

---

[3] http://www.science.aster.ersdac.or.jp/en/documnts/users_guide/(September 2005).

[4] http://www.science.aster.ersdac.or.jp/en/documnts/users_guide/(September 2005).

Fig. 7.15 Aster satellite scenes of the researched sites. Provided by EOS Data Gateway

Scaleparam.	Color	Shape	Compactness	Smoothness
25	0.7	0.3	0.5	0.5
50	0.7	0.3	0.5	0.5
75	0.7	0.3	0.5	0.5
100	0.7	0.3	0.5	0.5
150	0.7	0.3	0.5	0.5
200	0.7	0.3	0.5	0.5

Fig. 7.16 Segmented Aster satellite scenes

A purely pixel-based classification would have difficulties here in compiling a good classification model. It is also useful to include the elevation and slope information from the DEM in the classification project. In this way, mountains can be best separated from the other classes by a slope texture. A more detailed interpretation leads to the classification model shown in Fig. 7.17. All four ASTER satellite scenes can be classified on the basis of this model. Figure 7.18 shows the result of this classification.

The quality of the classification is fairly good. Starting from the goal of detecting **industrial sites** in large-scale satellite images, it must be remembered that all the facilities of interest were recognized by the model. Furthermore, misclassifications of industrial plants are found for all three satellite images since some objects were classified as such plant when this was not the case. However, apart from the training objects, the model also identified other industrial plants in the individual scenes.

In general, also the result for the other classes make a good impression. A quantitative quality assessment will not be performed for this case study. The reason is the small number of objects available for the individual classes. For the class of **industrial site** at Natanz only one object is identified and there are two misclassified objects, which would lead to a *user accuracy* of 33%. Although this is statistically correct it cannot be compared with the other quality estimates where far more reference objects are available as a basis. Furthermore, there are very large objects, which in many cases consist of mixed use, mainly for the classes of **soil** and **mountains**. A quantitative determination of quality would consequently not be comparable with the others in this study and would not be sufficiently statistically significant. Since, on the whole, all the facilities sought and also some new ones were identified and since there were not too many misclassifications for the class

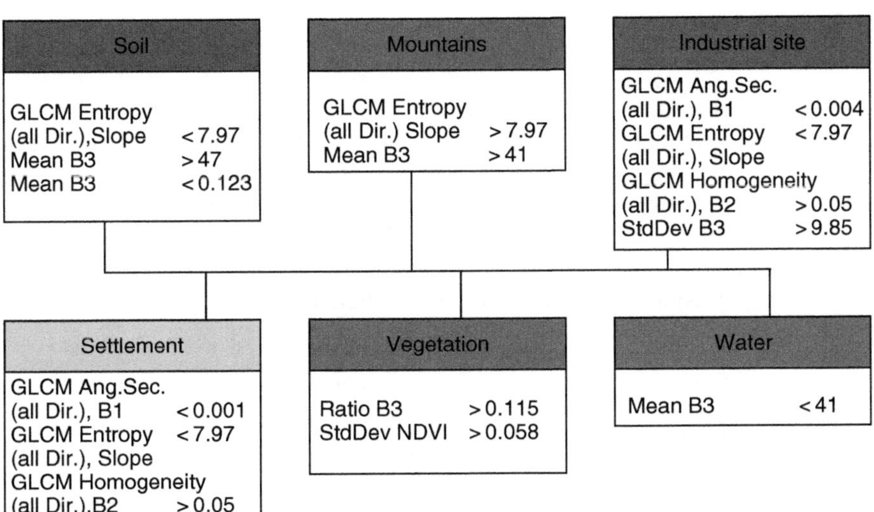

Fig. 7.17 Classification model of the Aster data

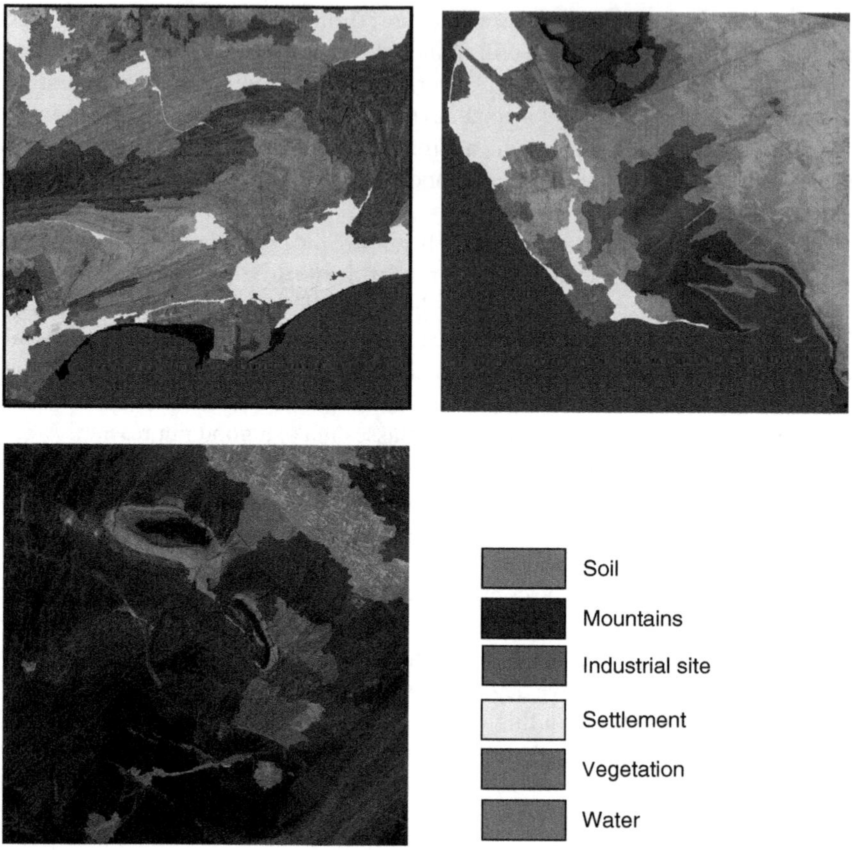

Fig. 7.18 Classification result of the Aster satellite data

of **industrial sites** the results of the model can be regarded as relatively good. It is suitable for detecting facilities in large-scale satellite scenes. The partial overlap with settlements or regions in mountainous areas which appear similar to the plants because of their texture, can be excluded fairly rapidly by using an interpreter.

The methodology itself is consequently suitable for making a preliminary selection of potential targets as part of a monitoring and verification process. However, due to the misclassifications, it cannot replace a human interpreter, but it can provide valuable preliminary work.

7.5 Bushehr Site – A Potential Outlook

Object-based image analysis has a very great potential for analysing satellite images in new and innovative ways. The eCognition software applied here has proved especially suitable. Since, however, the starting point for all results produced by object-based image analysis is the segmentation, the latter decisively influences the possibilities and limitations. Our studies presented in the previous chapters are all based on a standardized automated segmentation without manually improving the results of the object extraction. This illustrates the present state of the art and forms the basis for all current work. At the same time, however, this form of segmentation also limits the potential of the image analysis since the object boundaries identified do not exactly correspond to the actual objects.

Some features cannot therefore be used for describing the objects since the segments found do not adequately reflect the objects. One example of this is the reactor building, which is typically characterized by its round shape. However, with segmentation such a reactor building cannot be extracted as an exact round segment but rather takes on a distorted, fractal character. It is thus difficult to integrate existing knowledge-based features in the analysis since these known features were established for real world objects.

An analysis of the potential of the methods of object-based image analysis will be performed in this chapter with the example of the Iranian nuclear power plant in Bushehr, on the basis of objects that had been very well extracted. Furthermore, features from a-priori knowledge are included in the feature description of the object classes. This is based on the so-called *key features* in Jasani (2004). In this way, it should also be possible to semantically identify individual buildings within the facility. A basic requirement for implementing the key features is very realistic object boundaries since the key features mainly refer to shape parameters.

The basis of this analysis is a preprocessed QuickBird-2 satellite image of the Bushehr nuclear power plant, Iran, from 26 June 2003. This is shown in true colours in Fig. 7.19. The segmentation was performed here by adding a vector layer on the basis of which the boundaries of large buildings within the plant are detected. This layer was first manually generated in GIS software. If the vector layer is included in eCognition as a so-called thematic layer the object boundaries found on it will be taken into consideration in the subsequent segmentation. This results in very well defined buildings within the plant. The segmentation settings here are 0.3 for the *shape factor* and this is weighted in 0.5 *compactness* and 0.5 *smoothness*. The second level (scale parameter 100), the fourth level (250) and the sixth level (600) are used for modelling. Table 7.11 shows the segmentation.

In the sixth hierarchy level, the classes of **background**, **hybrid cooling towers**, **buildings**, **walls** and **watchtowers** are defined in a first step. The partial transferability of the model of NFRPC Esfahan (see Section 7.3) means that features can be taken over for some object classes. The other features are determined via a SEaTH analysis with 255 features.

Figure 7.20 shows the complete classification model. After general modelling of the buildings, certain building types were classified. The model will therefore be in a position to identify the **pressurized water reactors (PW NPS)**, the **turbine and generator hall**, as well as the reactor **control building**. These are defined as subclasses of the buildings class. A priori knowledge is now integrated into the analysis. The feature description of these classes corresponds to the key features

Fig. 7.19 QuickBird scene of Bushehr, June 2003. Original Data/Product © Quickbird/Digital Globe; 2003; Distributed by Eurimage

drawn up by Jasani (2004) on the basis of German pressurized water reactors. It will be examined whether these key features can also be applied here and thus whether it is possible to separate individual buildings.

Table 7.11 Parameters for the Bushehr segmentation

	Scale Parameter	Color	Shape	Smoothness	Compactness
Level 1	50	0.7	0.3	0.5	0.5
Level 2	100	0.7	0.3	0.5	0.5
Level 3	150	0.7	0.3	0.5	0.5
Level 4	250	0.7	0.3	0.5	0.5
Level 5	400	0.7	0.3	0.5	0.5
Level 6	600	0.7	0.3	0.5	0.5

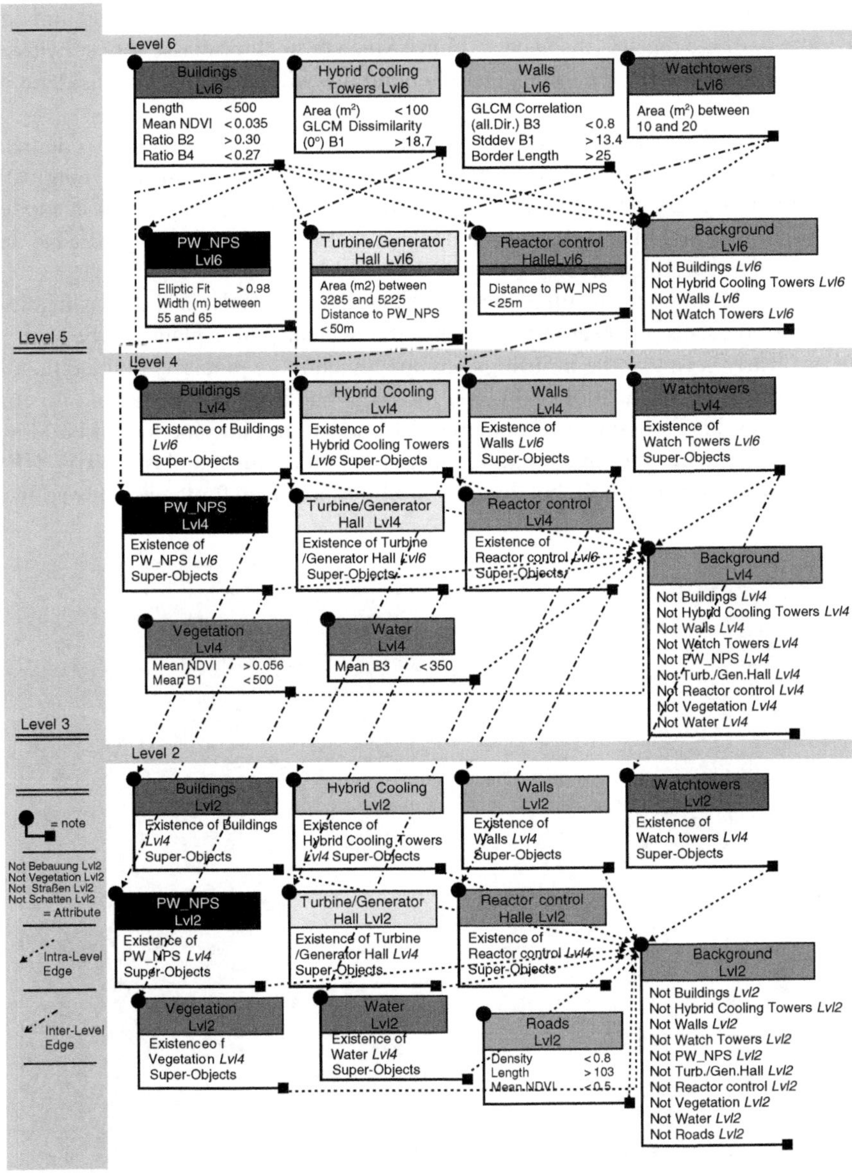

Fig. 7.20 Classification model for the scene Bushehr, July 2003

The pressurized water reactors are described by a round shape (*elliptic fit* > *0.98*) as well as by a certain diameter (width between 55 and 65 m). The turbine and generator hall is characterized by certain dimensions which permit it to be defined by a certain surface measurement (*area* between 3825 and 5225 m^2). Furthermore, this hall is located close to the reactor (*distance* < 50 m). There is no precise description for the reactor control building in Jasani (2004), only that it is located in the immediate vicinity of the reactor (distance < 25 m). Since these three classes are subclasses of the buildings class, it is certain that only buildings will be taken into consideration as objects.

At the fourth level, the classes of **vegetation** and **water** are then also defined. The class of **roads** is modelled on the second level. **Background** is dynamically determined at every level. Figure 7.20 shows the complete classification model, which is composed of transferred features, of features determined by SEaTH and of knowledge-based features.

Figure 7.21 shows the result of the object-based image analysis, the classification. The figure presents a subset in which only the nuclear power plant can be seen. It can be clearly seen that the **buildings** are now much more precisely defined than in all the previous classifications and only the large buildings will be taken into consideration in this assessment of potential. The separation into individual buildings can be seen, where only one building each is identified for the class of **PW NPS**, **turbine and generator hall**, but three possibilities are given for the **reactor control**

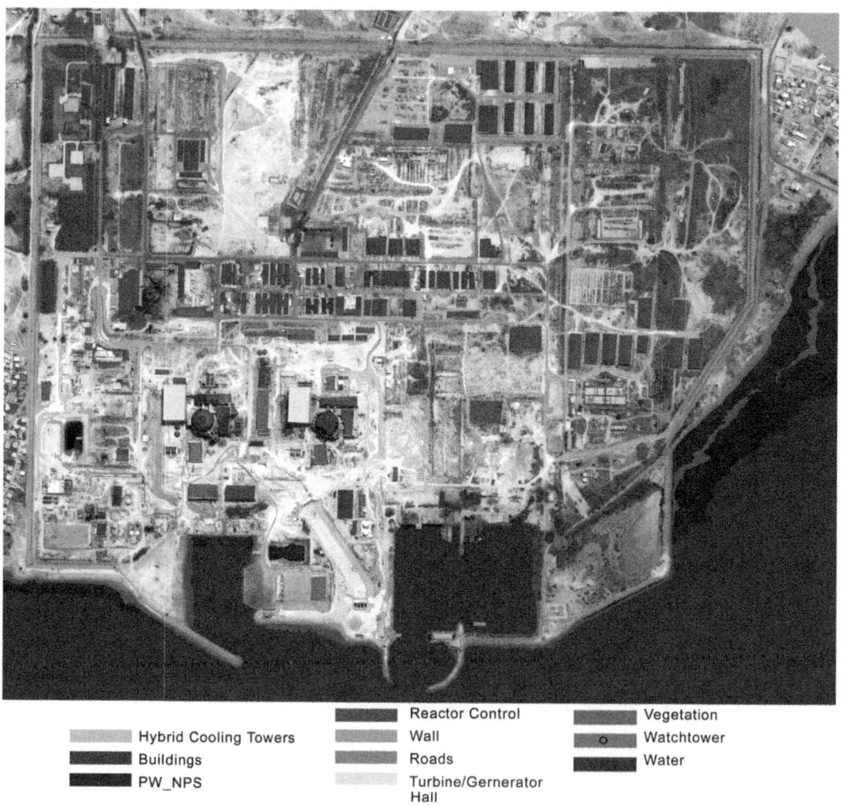

Fig. 7.21 Classification of the satellite scene for Bushehr, July 2003. Original Data/Product © Quickbird/Digital Globe; 2003; Distributed by Eurimage

building. This is due to the thin rule base for classifying the reactor control, which is based merely on vicinity to the reactor. Furthermore, **hybrid cooling towers**, a large **wall** around the facility, and a large number of **watchtowers** are also identified. The nonsensitive classes, such as **water**, **vegetation**, **roads** and the remaining buildings make a qualitatively good impression.

As in previous chapters, the quality of the classification of the plant is performed here quantitatively via the overall accuracy and the user and producer accuracy (cf. Section 7.1.3). The reference objects for the sensitive classes were visually determined by Jasani prior to this study. The reference objects of nonsensitive classes were also determined visually. The choice of correct reference objects was ensured by the high spatial resolution and consultation of an expert for the sensitive objects.

Table 7.12 summarizes the results of the quality assessment. For reasons of space, the reference classes are represented by numbers corresponding to the class numbers of the classified objects. The quality is determined on the second scale level so that large objects are divided here into smaller ones. Due to the selected classification model, the totality of the small objects always corresponds to the large superobject. The second level is selected for the assessment since all the object classes are represented here and it is thus possible to perform a holistic quality assessment.

For the sensitive classes of **walls**, **PW NPS**, **reactor control**, **turbine / generator hall**, **hybrid cooling towers** and **watchtowers**, all actually existing objects in these classes were selected as reference objects. The significance level of the quality assessment for these classes thus corresponds to 100%. This makes it possible to

Table 7.12 Qualitätsabschätzung der Klassifikation von Bushehr, Juni 2003

Classified Objects	Reference Objects										
	(1)	(2)	(3)	(4)	(5)	(6)	(7)	(8)	(9)	(10)	$\sum$
Roads (1)	28	0	0	0	0	0	0	0	0	0	28
Vegetation (2)	0	48	0	0	0	0	0	0	0	0	48
Water (3)	0	0	23	0	0	0	0	0	0	0	23
Walls (4)	0	0	0	6	0	0	0	0	0	0	6
Buildings (5)	0	0	0	0	34	0	0	0	1	0	35
PW_NPS (6)	0	0	0	0	0	9	0	0	0	0	9
Reactor C. (7)	0	0	0	0	4	0	6	0	0	0	10
Turb./- Gen. (8)	0	0	0	0	0	0	0	10	0	0	10
Hybrid CT (9)	0	0	0	0	0	0	0	0	7	0	7
Watchtowers (1 0)	0	0	0	0	0	0	0	0	0	10	10
unclassified	18	6	0	0	1	0	0	0	0	3	28
$\sum$	46	54	23	6	39	9	6	10	8	13	214
Producer Accuracy	0.47–0.75	0.81–0.97	1.00	1.00	0.77–0.97	1.00	1.00	1.00	0.88	0.77	
User Accuracy	1.00	1.00	1.00	1.00	0.91–1	1.00	0.60	1.00	1.00	1.00	
Overall Accuracy σ	0.02										
Overall Accuracy	0.85–0.93										

dispense with a calculation of the standard deviation. For the other classes, only some of the objects are selected as reference objects so that the quality assessment here is performed on the 95% significance level ($\pm 2\sigma$). For the calculation see Section 7.1.3. The "unclassified" row in the table represents the objects which cannot be assigned to any class and thus belong to the **background** class. No quality assessment was performed for these objects.

The two reactors (**PW NPS**) were recognized by the model and achieve a classification quality of 100% in all dimensions. The situation is the same for the classes of **walls** and **turbine/generator hall**. Of the 13 **watchtowers** present, 10 were correctly identified, which leads to a producer accuracy of 77%. More **reactor control buildings** were found than actually exist. This resulted in a user accuracy of "only" 60%. This statistically poor result is due to the inadequate knowledge-based feature basis for this class. The **hybrid cooling towers** achieve a producer accuracy of 88% because one cooling tower was not recognized. The classes of **buildings** and **vegetation** achieved good classification results, whereas there were a large number of misclassifications for **roads**. However, these misclassifications always lead to an assignment as **background** (unclassified). The overall quality of the classification is 85%–93% on the 95% significance level.

This image analysis displays the great potential of object-based, rule-based classification. On the basis of high-resolution images, it is possible to adequately identify even sensitive buildings. In this way, potential nuclear facilities can be distinguished from industrial sites for verification purposes under the NPT. The restriction is the present state of the art for segmentation. Although the identification of buildings can be fully automated with the methods examined in this study, it is difficult to recognize typical buildings in the nuclear fuel cycle. Only a manual preprocessing step adequately defines building objects so that they can then be further differentiated.

As an alternative to a manual intervention in the segmentation, it could be possible to develop an improved segmentation algorithm. It should then be possible on the basis of SEaTH or a priori knowledge to develop suitable reference models for various facilities in the nuclear fuel cycle. This can then be used to support the verification process under the NPT.

The following chapter on combined change detection is also concerned with the potential for NPT verification. To this end, a time series of satellite scenes of NFRPC Esfahan will be analysed. The aim here is the automated detection and classification of changes so that, in turn, there will be no need for manual intervention.

Chapter 8
Combined Change Detection at NFRPC Esfahan

This chapter is concerned with the issue of how multitemporal changes of individual objects within (nuclear) facilities can be automatically recognized and semantically determined. Although the existing pixel-based MAD method (multivariate alteration detection – see Section 4.2) is very well suited for detecting significant changes in satellite images, an interpretation or identification of the changes is only possible to a limited extent. In the present chapter, a new method of detecting **and** identifying multitemporal changes in very high-resolution RS data will be derived and assessed. First of all, the MAD technique will be applied to the multitemporal RS data from NFRPC Esfahan. Then, for the first time, a combination of pixel-based change detection and object-based image analysis will be developed and evaluated. The aim is to show that this combination has decisive advantages and that it represents a gain in information in comparison to conventional uncombined processes.

8.1 Combined Change Detection in the Period from 2002 to 2003

Data from the preprocessed satellite images (Chapter 6) of NFRPC Esfahan from 2002, 2003 and 2004 are available for multitemporal change detection. The aim is the detection of successive changes in the individual years. In the following, the period from 2002 to 2003 will be discussed in detail and studied with respect to changes. Furthermore, it will also be possible to identify the changes found by combining the object-based classification and the MAD change information. This period is taken as an example of the possibilities opened up for change detection by the approach selected here.

8.1.1 Pixel-Based Change Detection

The first change detection was performed for the period from 2002 to 2003. The MAD transformation is available as an open source IDL program on the Internet and can be implemented as a tool in the ENVI remote sensing software. The

© Springer Science+Business Media B.V. 2008

transformation resulting in the MAD change components proceeds fully automatically after the reference data have been selected.

Figure 8.1 shows the result of the MAD transformation. It presents the MAD components MAD2, MAD3 and MAD4 in a colour combination since the change signals of the four QuickBird channels are mainly concentrated in these colours. MAD 1 contains the greatest amount of noise and will not be used here. At the present point in time, no differentiation is made between significant and insignificant changes. The pixel value of each MAD component is its standard deviation σ from the mean of the respective MAD component distribution. A value of 0 means no change, the greater (+) or the smaller (−) σ the more probable a change becomes.

In Fig. 8.1, it can be seen that a lot of pixels have changed. No change would be marked by gray pixels. All other colours indicate changes. Over a period of one year it can be expected that almost every pixel has changed. It is consequently necessary to separate significant from insignificant changes.

Fig. 8.1 All the changes at NFRPC Esfahan in the period from 2002 to 2003

A differentiation into significant changes can be achieved by threshold values (cf. also Section 4.2). Since the MAD components have a normal distribution or are uncorrelated, a threshold value can be assigned via the standard deviation σ of the *no change pixels*. *No change* pixels represent the insignificant changes. The threshold value is determined separately for a predefined deviation from the mean for each MAD component. In order to obtain a significance level of 95% for the changes, the threshold is defined as 2σ of the *no change* pixels. If a pixel has a valued greater than 2σ then it is a *change pixel*. Table 8.1 shows the standard deviation calculated from the variance for MAD2, MAD3 and MAD4 of the *no change pixels*.

The threshold values determined in this way now indicate the significant changes in the period from 2002 to 2003. Insignificant changes, represented by *no change pixels* are shown in Fig. 8.2 in grey. There are significant changes in the region of the facility itself and also in the vegetation present. The surrounding hills and countryside display few significant changes.

Figure 8.2 provides information on where significant changes have taken place in the period from 2002 to 2003. An interpretation or an identification is, however, not possible on the basis of the MAD difference image. Initial indications are given by the nature of the individual pixel values of the MADs. Similar values in the MADs – shown in the picture by a similar colour – represent a similar change signal and thus possibly a similar type of change. A more detailed interpretation or a semantic determination of the changes is not possible on the basis of the MAD methodology alone. In the following, it will be shown that a combination of the pixel-based MAD technique and object-based classification has decisive advantages and new potential for interpreting the changes that have occurred.

8.1.2 Object-Based Change Classification

Object-based classification (cf. Chapter 7) and its combination with the change information from the MAD components can make a decisive contribution to interpreting the changes. The individual MAD components are integrated into the object-based image analysis as additional data. The use of eCognition makes this easy to do (cf. Chapter 3: keyword: multi-source aspect). In addition to the large number of object features, there are now also change features which can be used for object-based analysis.

Table 8.1 Covariance matrix of the no change pixels of the MAD components

Covariance Matrix of MADs for the *No Change Pixels*			
1.494	−0.0071	0.0196	0.0029
−0.0071	**0.4881**	−0.285	−0.0044
0.0196	−0.02885	**0.188**	−0.0647
0.0029	−0.0044	−0.0013	**0.0647**
MAD 1	**MAD 2**	**MAD 3**	**MAD 4**
Sigma	0.69	0.43	0.25
2 Sigma	1.4	0.87	0.5

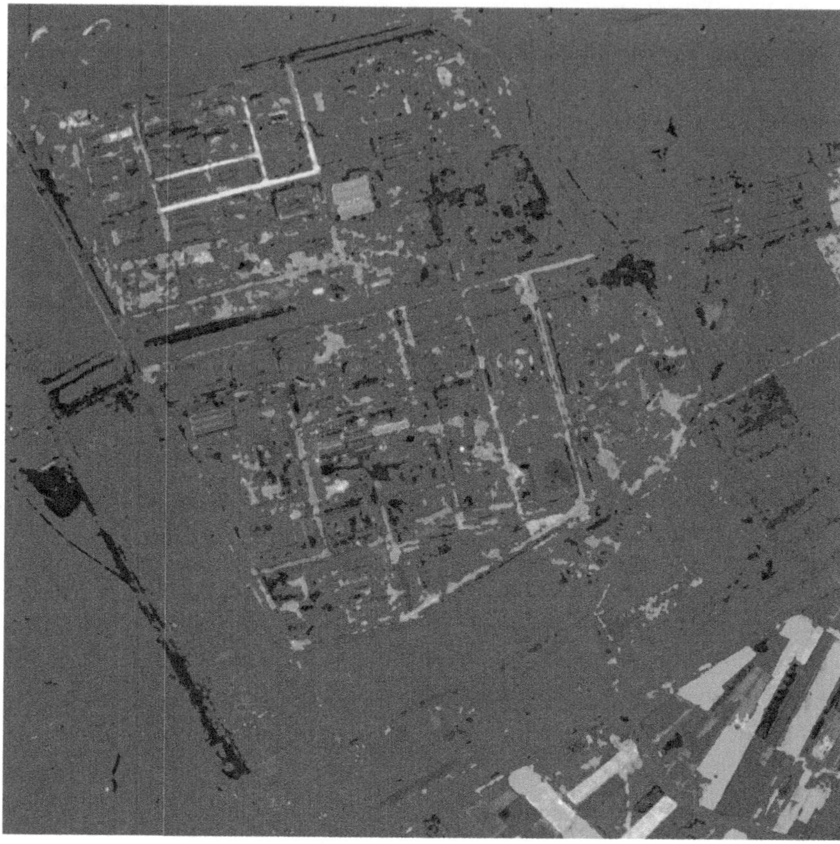

Fig. 8.2 Significant changes at NFRPC Esfahan in the period from 2002 to 2003

The reference classification model (cf. Fig. 7.3) developed with the SEaTH methodology (Chapter 5) from Section 7.1.3 will continue to be used for the semantic determination of objects. As already indicated, this can be transferred readily and in a technically simple manner to the year 2003 and the slight modification in the NDVI threshold made there is also retained. A more extensive classification of the changes can be derived from this by taking into consideration the change features of the objects. Only significant changes are taken into consideration, which are also determined via a threshold value of 2σ of the *no change pixels*.

Apart from the semantics of the objects, the classification now also differentiates between significant and insignificant changes. For each object of an object class it can thus be determine whether it has changed significantly in the period from 2002 to 2003 or not. Implementation in eCognition is performed on the basis of so-called *inheritance* (cf. Section 3.3). The parent classes are the classes of the semantic model i.e. **buildings, background, walls, fences and pipes systems, shadows, roads** and **vegetation**. These classes pass on their features to the respective child classes of **change** and **no change**. No change-classification is performed for the **shadows** class since due to a different camera angle the shadows had to be changed completely. If for an object there is a significant change signal in at least one of the 3 MAD components (MAD2, MAD3, MAD4) it is classified as a changed object, otherwise it is not. Figure 8.3 illustrates the classification model established.

When the model is applied to the preprocessed image of NFRPC Esfahan 2003 the change classification shown in Fig. 8.4 is obtained. Due to clarity reasons only the objects of the changed classes are shown. It is apparent that by linking object-based classification and pixel-based change detection it is possible to rapidly obtain a semantic classification of the changes. A good interpretation and identification of

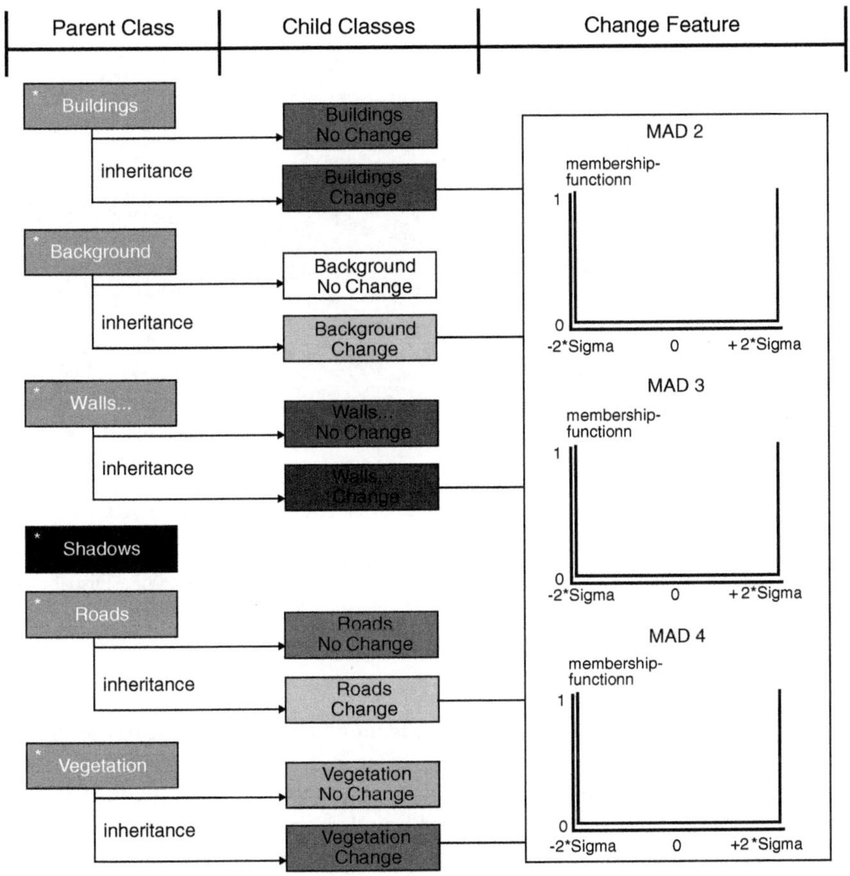

Fig. 8.3 Object-based classification model of the significant changes at NFRPC Esfahan in the period from 2002 to 2003. The parent classes of buildings, background, walls . . ., roads and vegetation are defined by the reference model in Section 7.2. An additional classification differentiates between change and no change. All image objects are regarded as significant change that have a change signal in one of the three MADs $\pm 2\sigma$

Fig. 8.4 Change classification of the significant changes at NFRPC Esfahan in the period from 2002 to 2003. Original Data/Product © Quickbird/Digital Globe; 2002–2003; Distributed by Eurimage

the changes is thus possible since the significant changes can be clearly assigned to the object classes.

There have been changes in all classes. As expected, the **background** class had the least number of significant changes. Although some change objects are assigned to the **background** class, this semantics is not always correct. The misclassification is understandably based on the quality of the classification model used. In this way, for 2003 the **background** class achieves a *producer's accuracy* of 67%–85% on the 95% significance level, which reflects just this misclassification. The **vegetation** class displays a large number of changes which can be found in the plantations around the NFRPC and the fields at the bottom edge of the picture. There are some changes with the **roads**, and in the **buildings** class changes have taken place above all in the northern part of the facility.

The present level of information now makes it possible to establish where and in which class changes have taken place. This is a significant increase in the amount of information obtained in comparison to purely pixel-based change detection. The methodology is consequently in a position to specifically identify changes in certain classes. For the verification of the NPT, for example, changes in the buildings sector could be of great significance. Since, however, different changes may take place in the individual classes, a differentiation into various types of changes within the object classes would be desirable. This issue will be discussed in detail in the following.

8.1.3 Class-Related Change Determination

This section is concerned with the different types of change within a class. Thus, for example, in the **buildings** class there may be changes due to existing buildings or changes may occur due to the construction of new buildings. An attempt will be made in the following to differentiate these changes within a class. The potential for making this differentiation can be found in the nature of the MAD components. As already mentioned in Section 8.1.1, a similar change in the individual MAD components also means a similar type of change. This information can be used to obtain an in-depth interpretation.

In order to consider individual classes separately, vector data sets for the individual object classes are generated in eCognition. They can then be considered separately in a GIS. The advantage of this methodology is the possibility of focusing on just one class. In this way, for example, road construction activities at a facility can be observed over a considerable period of time. The same is true of the other classes. The classes of **buildings** and **roads** were analysed in order to investigate the potential of this methodology. Figure 8.5 shows the vector data sets for these classes.

The aim of the following classification is to discover whether, depending on the MAD component, other types of change can be differentiated. Beginning with the **buildings** class, a more extensive classification model will be established for this class. The different change features in the MADs can be divided into classes. First of all, a positive and a negative direction of change ($+2\sigma$ and -2σ) must be distinguished. Furthermore, there are three MAD components here so that a positive and negative change signal is obtained for each of them. Consequently, six classes are modelled. Each class represents the strongest change signal of a MAD component in a positive and negative direction, respectively.

In order to identify the strongest change signal for an object, the individual values of the MADs must be comparable with each other. This can be achieved by normalizing the individual components. To this end, the values of the MADs are divided by their respective standard deviation. The normalization means that a standard deviation of 2σ corresponds to a MAD value of 2 in every component.

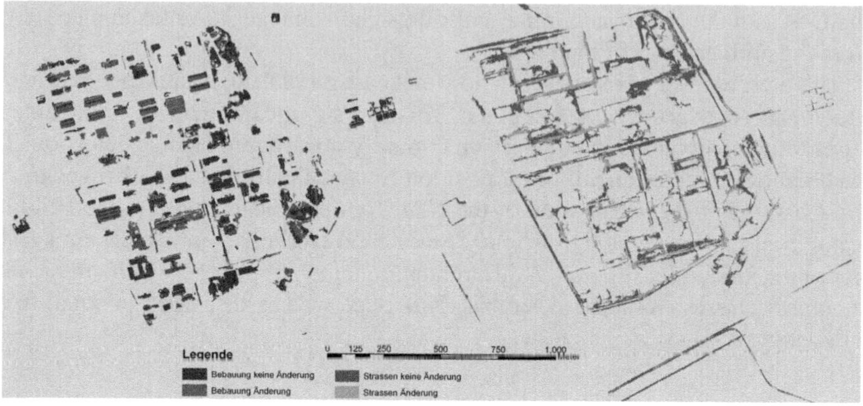

Fig. 8.5 Vector data of the change classes of buildings and roads

By selecting an ascending membership function in eCognition it is now possible to identify those MAD components with the highest degree of membership, that is to say the strongest change signal. Sixteen standard deviations are laid down as the maximum value. Figure 8.6 illustrates diagrammatically the classification model for the **buildings** class.

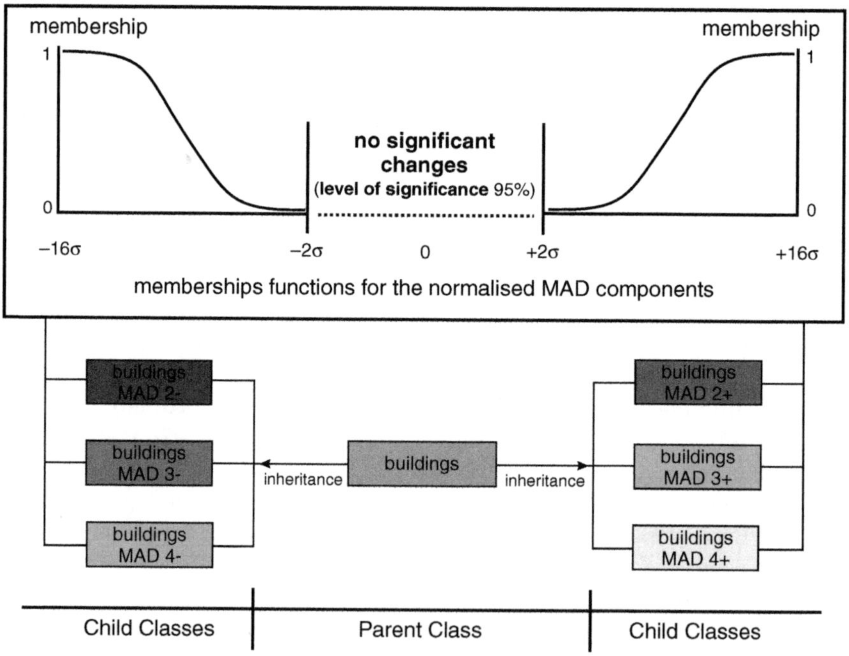

Fig. 8.6 MAD-based classification model of the significant changes in the **buildings** class at NFRPC Esfahan in the period from 2002 to 2003

Figure 8.7 shows the result of the MAD-based object-based classification with this model. The different types of change can be identified very clearly here. The two dominant classes are **buildings MAD2+** and **buildings MAD3+**. The other classes are represented to a lesser extent or hardly at all. The dominant change classes of buildings are considered in more detail in Fig. 8.8 in comparative image subsets. The **buildings MAD2+** class thus apparently mirror the reflectance changes of the roofs of the buildings. The significant reflectance changes result from the different angle of insolation at the time the image was recorded in 2003. The **buildings MAD3+**

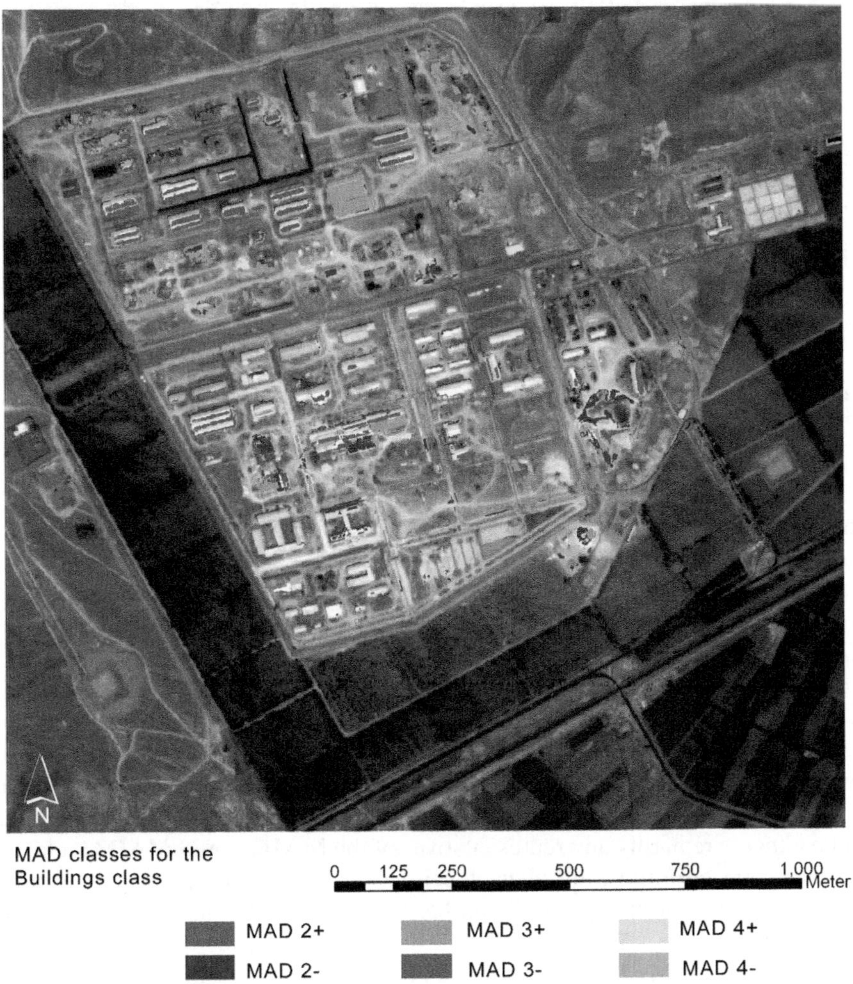

Fig. 8.7 Result of MAD-based classification of significant changes in the **buildings** class at NFRPC Esfahan in the period from 2002 to 2003. Original Data/Product © Quickbird/Digital Globe; 2002–2003; Distributed by Eurimage

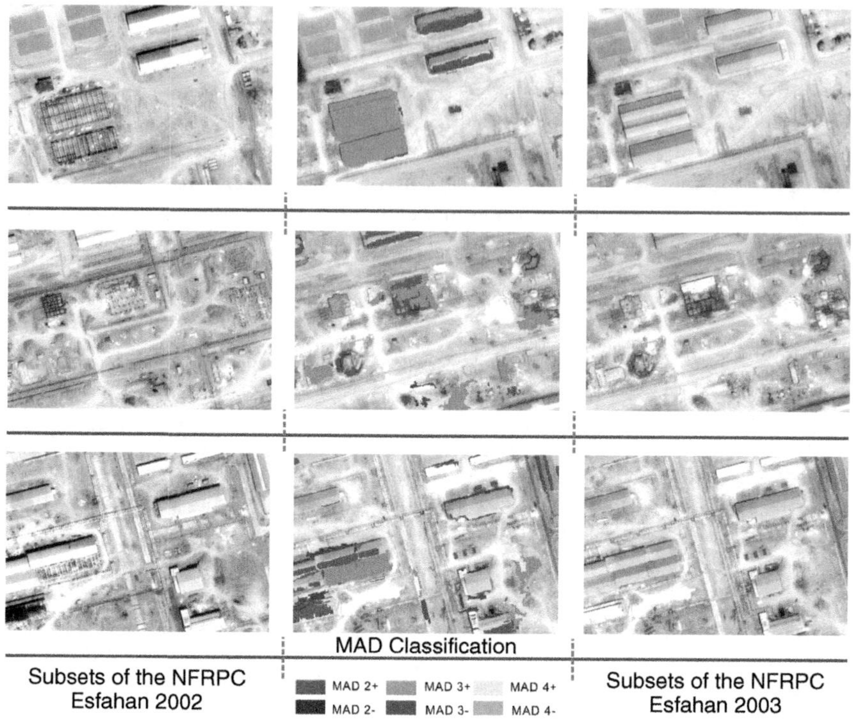

Fig. 8.8 Subsets of the MAD-based classification of significant changes in the **buildings** class. Original Data/Product © Quickbird/Digital Globe; 2002–2003; Distributed by Eurimage

class can therefore be used to verify the NPT. These changes are thus mainly newly erected building structures.

It can be seen that buildings that were not yet, or only partially, erected in 2002 but were completely or partly finished in 2003 can be identified in the **buildings MAD3+** class. This classification therefore provides information on the type of change with respect to buildings and newly erected building structures can also be recognized. There are only a few representatives of the classes of the fourth MAD component, which points to the fact that these classes show changes other than buildings. A MAD-based classification of the semantic class of **vegetation** shows that the **MAD4±** components reflect changes in the vegetation sector. In this context, there are hardly any representatives of the **MAD2± and MAD3±** classes.

Apart from **buildings**, all other object classes can also be classified according to different types of change. This can be done very easily and rapidly within the eCognition software. This shows the innovative nature of the linked classification approach since the MAD child classes can be "attributed" to a different parent class without difficulty. This can be done in less than a minute so that a wide range of additional information for interpretation purposes can be generated in a short time. Figure 8.9 shows the change classification with the example of the **roads** class. As in

Fig. 8.9 MAD-based classification of significant changes in the **roads** class at NFRPC Esfahan in the period from 2002 to 2003. Original Data/Product © Quickbird/Digital Globe; 2002–2003; Distributed by Eurimage

the buildings class, the changes are predominately in the **MAD2+** and **MAD3+** classes here as well. By enlarging the subsets in Fig. 8.10, it becomes clear that both change signals indicate newly constructed roads, which, however, differ in the type of road surfacing.

The possibilities discussed above of automatic change detection, classification and interpretation help to make the image analyst's work easier and they can be used to make a preliminary selection of targets that are to be considered. When handling large volumes of data, as arise with safeguards and verification of nuclear facilities, this can lead to considerable savings in time and thus money. It must, of

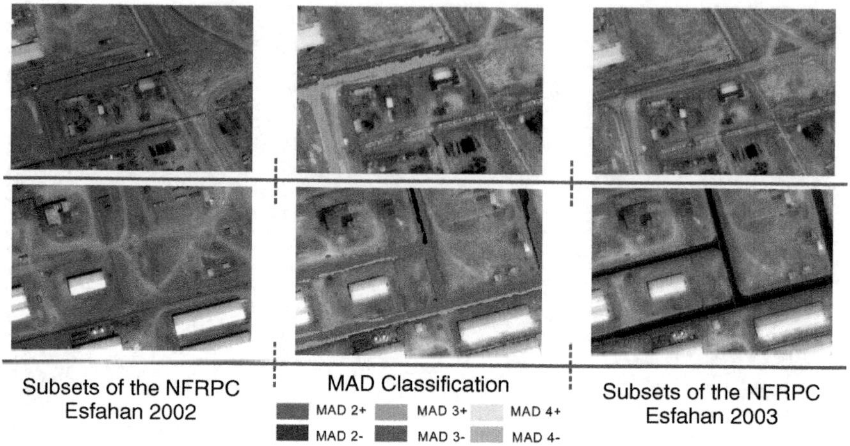

Fig. 8.10 Subsets of the MAD-based classification of significant changes in the **roads** class. Original Data/Product © Quickbird/Digital Globe; 2002–2003; Distributed by Eurimage

course, be remembered that the classification can only be as good as the model it is based on. However, in view of the paucity of preliminary information available on the area under investigation and the use of commercial satellite data this represents a considerable gain in information and interpretation. The existing models can, of course, be continuously improved and extended.

8.2 Combined Change Detection in the Period from 2003 to 2004

The period from 2003 to 2004 will only be discussed briefly and summarized in the following. There are several reasons for this. Firstly, the work flow is identical to that for the years 2002–2003, which has already been discussed in detail. The other reasons are conditioned by the data material available. The scene recorded in May 2004 differs quite considerable with respect to its in-track and cross-track camera angle. Consequently, when the buildings and other high structures are being registered during preprocessing (Section 6.1) it is not possible to achieve precise agreement. This leads to a large number of changes due to misregistrations of the image data indicating different positions for identical structures. Here a orthorectification of the data could be useful (cf. Section 6.1). Improved RS data (comparable camera angle) would undoubtedly be desirable, but is not available for the period in question. This indicates a restriction with respect to change detection for high-resolution satellite images. The archives of commercial suppliers contain very few (in some cases just one single) picture(s) for the period under consideration. The reason for this is the coverage of relatively small areas per scene and in the case of all optical images it is rarely the case that the scene is free of cloud cover.

The problem of misregistration for both sets of image data is illustrated in Fig. 8.11. Here stripes of the automatically recorded 2004 satellite scene are su-

Fig. 8.11 Misregistration of scenes at NFRPC 2003 and 2004. Original Data/Product © Quickbird/Digital Globe; 2003–2004; Distributed by Eurimage

perimposed on the 2003 image of the NFRPC. The black arrows draw attention to the pixel inaccuracies. It is difficult to perform change detection on the basis of such data material. It must be expected that there will be a large number of false alarm changes, which may additionally conceal real changes. Nevertheless, an example of pixel-based change detection will be performed here with a subsequent object-based change classification. Class-related change determination could help to separate the changes caused by misregistration from real changes and thus to ultimately also identify newly erected buildings.

A detailed derivation of the individual steps in pixel-based change detection and object-based change classification will not be presented here since these steps correspond to those performed in Sections 8.1.1–8.1.3. The classification model shown in Fig. 8.3 will also be used for change classification. The result of the change detection and subsequent change classification is shown in Fig. 8.12.

It becomes clear that many changes to **buildings** have taken place. These changes are often located at the bottom right-hand edge of the buildings. This accurately reflects the changes caused by misregistration. Further significant changes have taken place with the **roads** and **vegetation**. The figure shows all changes greater than 2σ, that is to say those that are on the 95% significance level. In order to distinguish changes due to misregistration from real changes, the **buildings** class will now be considered in more detail. Starting from the assumption that newly erected buildings will have a very strong change signal, only those changes in the buildings sector will now be shown whose change signal deviates from the mean by more than 10σ. Fig. 8.13 shows the classification model for the MAD-based change determination of the **buildings** class.

Fig. 8.12 Change classification of the significant changes at NFRPC Esfahan in the period from 2003 to 2004. Original Data/Product © Quickbird/Digital Globe; 2003–2004; Distributed by Eurimage

As an illustration, a section of the overall scene from June 2004 will be considered. It becomes clear that strong significant changes only exist for the **buildings** class in the MAD classes **MAD2−**, **MAD3+** and **MAD4−**. Only the class is shown in which the object had the strongest change signal. In the case of buildings, the main focus is on the **MAD4−** class.

A membership analysis of the individual buildings objects indicates that some buildings have a very strong change signal (and thus also a high level of membership) in only one of the change components (MAD2−, MAD3+ or MAD4−), whereas others have a strong signal in two change components (MAD3+ and MAD4−). Figure 8.14 shows the classification according to strongly significant changes with membership values of some building objects.

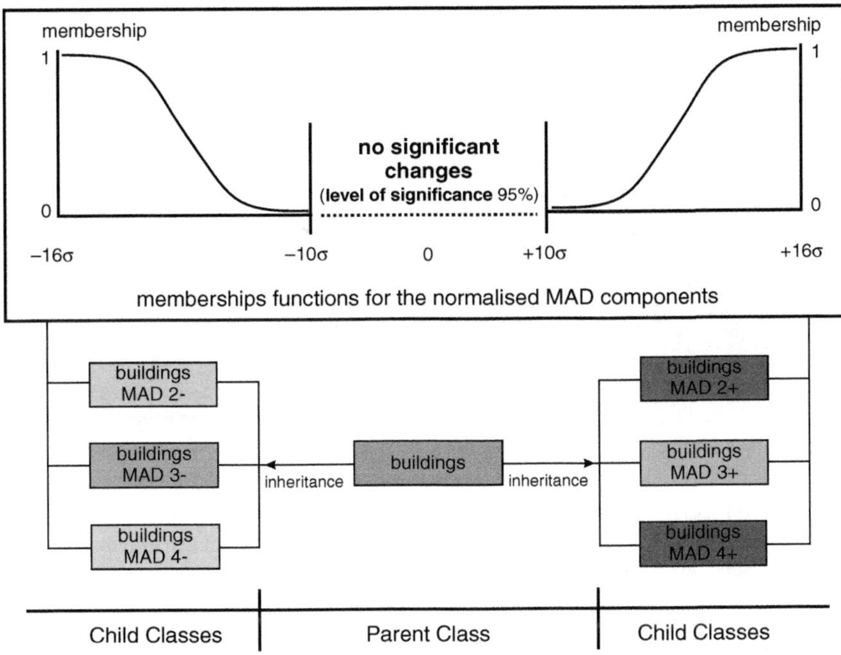

Fig. 8.13 MAD-based classification model of the strong significant changes in the **buildings** class at NFRPC Esfahan in the period from 2003 to 2004

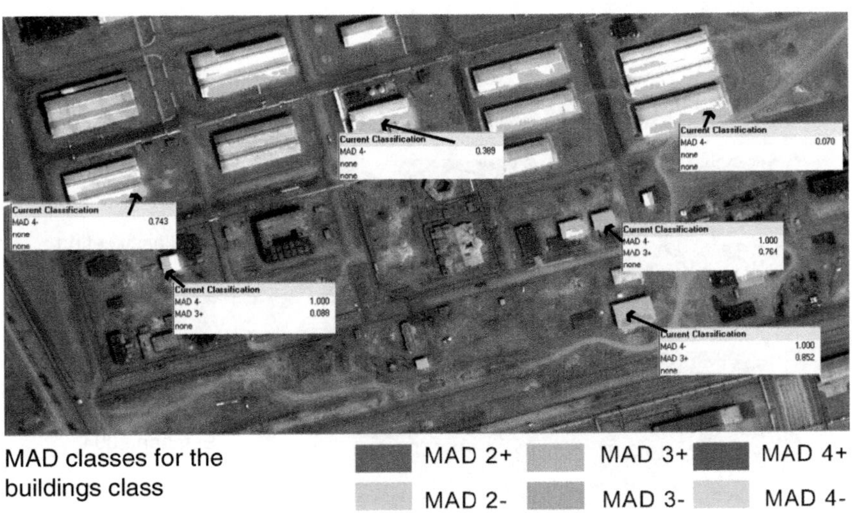

Fig. 8.14 MAD-based classification of the strong significant changes in the buildings class at NFRPC Esfahan in the period from 2003 to 2004. Original Data/Product © Quickbird/Digital Globe; 2003–2004; Distributed by Eurimage

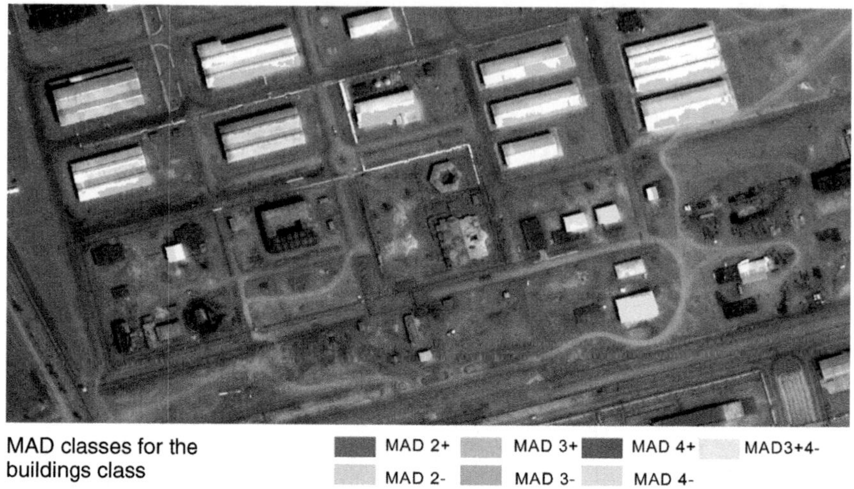

Fig. 8.15 MAD-based classification of the strong significant changes with combination class MAD3+4− at NFRPC Esfahan in the period from 2003 to 2004. Original Data/Product © Quickbird/Digital Globe; 2003–2004; Distributed by Eurimage

An additional class, **MAD3+4−**, is introduced in Fig. 8.15. This class identifies building objects which have a strongly significant change signal in both MAD components (MAD3+ and MAD4−). It can now be seen that some objects clearly stand out against the others. The qualitative comparison in Fig. 8.16 makes it clear that

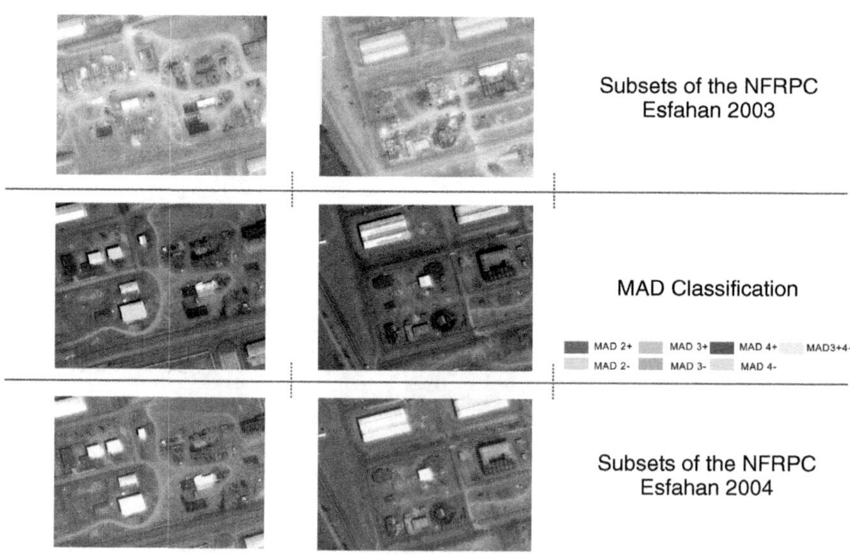

Fig. 8.16 Subsets of the MAD-based classification of the strong significant changes for the **buildings** class. Original Data/Product © Quickbird/Digital Globe; 2003–2004; Distributed by Eurimage

these are buildings that have been newly erected or completed in the period from 2003 to 2004.

These studies, performed as an example, have shown that in spite of poor data quality, newly erected buildings can also be identified by the selected methodology over a period of one year. Nevertheless, a greater number of steps are required than in the case of good-quality data (Section 8.1). Such steps can, however, be performed rapidly and without complications by the eCognition software. The extent to which these steps may later be automated is a subject for further study. However, the object of the development is to provide assistance for human users in their work and not to replace them entirely.

Chapter 9
Summary and Conclusions

An automated preprocessing of very high-resolution RS data makes new demands on the methods and algorithms used.

In registering individual satellite images the usual procedure of making a manual search for suitable TPs is made more difficult by the very high resolution. Furthermore, the time-consuming manual setting of TPs defeats the object of automating the working process. Panchromatic sharpening of the data requires new algorithms which optically sharpen the satellite images very well while at the same time modifying the spectral values of the multispectral data as little as possible. Older techniques, such as sharpening via the IHS method, considerably distort the spectral values of the high-resolution data so that these procedures are not appropriate for very high-resolution data. An automated relative radiometric normalization of multitemporal data always involves the difficulty of identifying suitable time-invariant regions of the individual satellite images for normalization purposes and this difficulty increases with increasing resolution of the data used.

Against the background of these requirements and the long-term aim of automated verification of the NPT, preprocessing steps are discussed in Chapter 6 and evaluated with respect to their applicability for very high-resolution satellite data. Most of these methods have been developed within the framework of project work at Research Centre Jülich and in cooperation with the GMOSS expert network.

The method used for geometrical correction of the RS data, image-to-image registration (cf. Section 6.1), is semiautomatic and provides very good results for the initial data. It has been demonstrated that more than enough TPs can be found for registration even with the high temporal distance and spatial resolution of the RS data to be processed. The method applied enables very high-resolution RS data to be registered fast and (semi-)automatically. It is furthermore also possible to make use of these TPs for all other registration algorithms. In future, the registration routine could be made fully automatic by improved programming and the inclusion of existing georeferencing.

In the present study, it was not possible to further investigate the orthorectification of high-resolution RS data. The reason for this was the lack of information on the terrain and high-resolution DEMs or DSMs as a basis for the orthorectification. Nevertheless, it has been demonstrated that if data material of sufficient quality

© Springer Science+Business Media B.V. 2008

is available a lack of orthorectification need not be a disadvantage. In the present study, it was possible to achieve good results by using the methodology of combined change detection developed here even without an orthorectification step (cf. Chapter 8). However, if the differences in the camera angle recording the RS data are too great and if high buildings etc. are present then this makes change detection without orthorectification more difficult (cf. Section 8.2).

Four algorithms were applied and assessed for the panchromatic sharpening of the RS data (cf. Section 6.2), two of which have been developed as part of project work at Research Centre J—lich. The aim of all the methods is a high level of detail sharpness with maximum spectral authenticity of the results. The studies performed have shown that there are differences in the quality of the results obtained by these more recent methods with respect to the targets set.

Daubechies wavelet provided the best results for spectral authenticity with the RS data used in this study (cf. Section 6.2.5 and Table 6.2). However, due to its poor performance for detail sharpness (cf. Fig. 6.11) this method will not be taken into consideration for further use. A good alternative is provided by the à trous cubic spline filter. In the quality assessment of spectral authenticity this method provided only slightly poorer values than that of the Daubechies wavelet and produced excellent detail sharpness.

Changes in reflectance in the image data to be investigated resulting from conditions at the time of recording can be corrected by radiometric normalization (cf. Section 6.3). The relative radiometric correction applied in the present study is fully automated and provided good results for data normalization. This is also an important prerequisite for analyses of the spatial and temporal transferability of reference models (cf. Chapter 7), whose good results demonstrate the quality of radiometric normalization.

Before actually applying object-based image analysis, a new algorithm was developed in the present study for the field of feature analysis. The analysis tool SeaTH, developed and validated in Chapter 5, is a first approach towards a comprehensive feature analysis as part of object-based image analysis. With SEaTH it is possible to extract characteristic features plus threshold values for any number of object classes from the large number of object features present. The result of the SEaTH analysis, the separability, is at the same time a measure of relative quality for the features investigated. A decisive part of the object-based image analysis, feature extraction, has now been satisfactorily solved on the basis of SEaTH. In this way, depending on the data material available, use can be made of a priori knowledge, SEaTH or a reference model, or a combination of all three, for a feature analysis.

When applied to the RS data to be analysed, SEaTH without exception provided good results. Prerequisites for the application of SEaTH for feature analysis are the availability of representative training objects and an approximately normal distribution of the probabilities of the features analysed. Since these conditions are not always found, SEaTH cannot be applied for feature analysis in all cases.

A major goal of the present study was to develop, apply and evaluate a new object-based methodology for automated image analysis in order to selectively obtain information for verification of the NPT from very high-resolution RS data. Furthermore, consideration was given to the extraction of typical features for the objects studied (nuclear facilities) and the establishment of reference models for object-based classification and the temporal and spatial transferability of those models. The quality of the results achieved was quantitatively analysed in order to evaluate the methodology.

The methodology developed in Chapter 7 and applied to RS data for nuclear facilities in Iran includes all the necessary steps for object-based image analysis, such as preprocessing, segmentation, feature analysis, semantic modelling, classification and evaluation of the results.

The multiscale segmentation of the RS data for object extraction is standardized in Section 7.1.1 for all subsequent investigations. The segmentation method selected achieves a good delimitation of all real-world objects in the segmentation levels generated. The parameters have to be standardized in order to create a common basis for all data sets and their further processing. This is of decisive importance for automating the process and for an objective comparative quality assessment of the results obtained.

Semantic modelling with the establishment of a rule-based classification model can be readily achieved by the feature analysis program SEaTH developed in this study. The classification model is created as a rule-based semantic network (cf. Section 7.1.2). The hierarchical structure of the classification model (cf. Fig. 7.3) enables the identification of heterogeneous real-world objects of different sizes and provides without exception good classification results. A comparison with a pixel-based classification showed that the quality standards achieved are far lower than those of an object-based classification. This is especially true of the classes of **roads** and **walls**. . ., which in high-resolution RS data can no longer be identified by spectral features alone. This demonstrates the superiority of a broader feature basis that also includes texture and shape features.

The temporal transferability of classification models for fast information acquisition has been examined and assessed on the basis of the RS data for NFRPC Esfahan 2003 (Section 7.2.1) and 2004 (Section 7.2.2). It has been shown that the features identified by SEaTH are characteristic of the individual object classes and can also be transferred over considerable periods of time. It was found that the features can be taken over completely and the threshold values can be used with minor adaptations to the reference model. The quality values achieved are good for the year 2003 though only satisfactory for 2004. Nevertheless, the temporal transferability is useful for obtaining rapid initial information from satellite images.

On the basis of various test regions, the spatial transferability of a reference model was verified (cf. Section 7.3). As in the case of temporal transferability, it became apparent that the features themselves are well suited for describing the classes, but that the thresholds in general, and in particular those of the spectral features, require adaptation. A spatial transfer of the reference model leads, on the whole, to unexpectedly good results and initial information can be obtained rapidly and simply from the satellite image by a minor adaptation of the threshold values. However, it is not possible to obtain more than a rapid overview and initial information from the spatially transferred reference model. In the case of different areas, the little more time-consuming SEaTH method leads to considerably improved results. The application of a reference model can, nevertheless, be useful if it is not possible to make a SEaTH analysis or in order to generate a basic framework for a classification model.

The object-based methodology developed for selectively obtaining information from very high-resolution RS data was applied and evaluated in Section 7.4 with respect to its transferability to low-resolution data. The first step in object-based image analysis, standardized segmentation, delimits the nuclear facilities (the objects to be identified) as image objects each in the Aster data for the Iranian sites of Arak, Bandar Abbas, Bushehr and Natanz. The feature analysis with SeaTH, making use of additional features derived from a DEM, leads to the establishment of a classification model. As a result, this reference model correctly identifies all the nuclear facilities sought on the four satellite images. In individual cases overlap arises with settlements or regions in the mountains which, due to their texture, appear similar to the facilities. These objects can be eliminated fairly rapidly by an image analyst. The methodology is consequently suitable for detecting industrial (nuclear) facilities in low-resolution data and thus of making a preliminary selection of potential targets as part of a monitoring and verification process. However due to the possibility of misclassifications this method cannot completely replace a human image interpreter.

The eCognition software used, in particular, and object-based image analysis, in general, opens up innovative ways of analysing satellite images and has very great potential for future applications. Since, however, the starting point for all results produced by object-based image analysis is the segmentation, the latter decisively influences the possibilities and limitations. At the moment the automated segmentation algorithm used in the eCognition software limits the general potential of the image analysis since the segmented object boundaries do not exactly correspond to the real world objects. At the moment no better algorithm is available for remote sensing (cf. Section 2.2). The segmentation can be improved by manual interventions. However, this does not make sense in view of the goal of an automated long-term verification process. In order to demonstrate the future potential of object-based image analysis with improved segmentation, the segmentation is improved manually in Section 7.5. This makes it possible, amongst other aspects, to include a priori knowledge in the existing methodology. With the aid of high-resolution satellite data it is now even possible to correctly identify individual building types (cf. Fig. 7.21). In this way, potential nuclear facilities can be distinguished from other industrial plant in order to verify the NPT.

The restriction in this case remains the present state of the art with respect to segmentation. Although the identification of buildings can be fully automated with the methods examined in this study, it is difficult to recognize typical buildings in the nuclear fuel cycle. Only a manual preprocessing step permits a very good delineation of building objects so that they can then be further diversified. If the quality of segmentation is improved in future then it should be possible on the basis of SEaTH or a priori knowledge to develop suitable reference models for various facilities in the nuclear fuel cycle. These models can then be used to support the NPT verification process.

Another objective of this study was to establish a methodology for automatically detecting and identifying changes in multitemporal data sets of very high-resolution RS data. The method of combined change detection, developed, applied and assessed in Chapter 8, is especially suitable for automatically detecting and semantically determining multitemporal changes of individual objects within (nuclear) facilities. A combination of the proven pixel-based MAD procedure and object-based image analysis thus achieves significant gains in information and interpretation as part of multitemporal change detection. This method makes it possible to automatically detect and identify changes and also to differentiate various types of change within individual object classes.

The potential indicated for combined change detection makes the work of an image analyst easier and a preselection can be made of the targets that need to be taken into consideration. When handling large volumes of data, as arise with safeguards and verification of nuclear facilities, this can lead to considerable savings in time and thus money. In view of the paucity of preliminary information available on the area under investigation and the use of commercial satellite data this represents a considerable gain in information and interpretation.

In summary, the potential of an object-based analysis method for supporting NPT verification with the example of Iran was to be assessed. Can the methods developed and applied here contribute to object-based image analysis for NPT verification? This question can on the whole be answered in the affirmative. Even though some further developments are desirable, such as an improvement of the quality of segmentation, even today the methods applied can provide a considerable gain in information. The methodology developed here is capable of analysing satellite images relatively fast and automatically and of identifying objects in nuclear facilities. Maps with compact information can be drawn up easily and rapidly on the basis of the results. Figure 9.1 shows an example of such a map. The maps shown here concentrate on the **buildings** class, but it is of course possible to generate maps for all other classes.

An image analyst can evaluate the satellite image visually and has as an additional information identified buildings highlighted. At more than 90% correct, the classification is very reliable. This can thus reduce the number of targets possibly overlooked by the analyst. However, it becomes obvious that a purely computer-based system is not sufficient at the present time. First of all, although the quality of the classification result of more than 90% correct is very good, for a verification of the NPT it should be close to 100%, at least for certain objects. The producer's accuracy of 100% given in Fig. 9.1 should also be considered in more detail. Statistically, the quality of the classification for buildings is 100% since all reference objects are correctly classified. However, this does not rule out the possibility that some existing building structures have not been identified by the model. Nevertheless, such a high level of classification quality leads one to expect that this could only affect a very few objects. The number that would later be regarded as acceptable cannot be decided here.

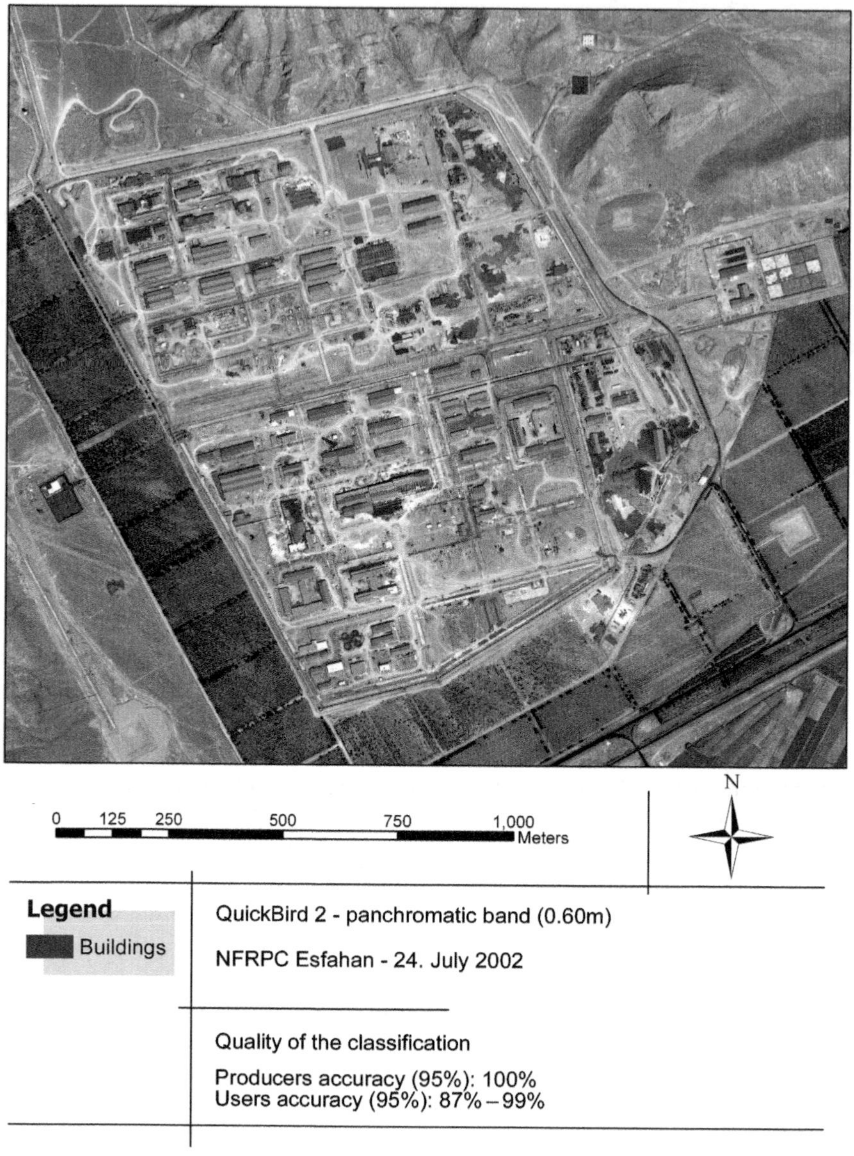

Fig. 9.1 Satellite image of the NFRPC Esfahan July 2002 with identified buildings. Original Data/Product © Quickbird/Digital Globe; 2002; Distributed by Eurimage

On the whole, a combination of computer-controlled and human verification approaches can be regarded as the most reliable method. These approaches can complement each other. The computer is capable of generating a large proportion of the information in advance and, on the other hand, of considerable reducing the volumes of data. Another possibility is to only analyse certain regions within a facility and thus to exploit the benefit of the very high spatial resolution of the satellite images used (cf. Fig. 9.2).

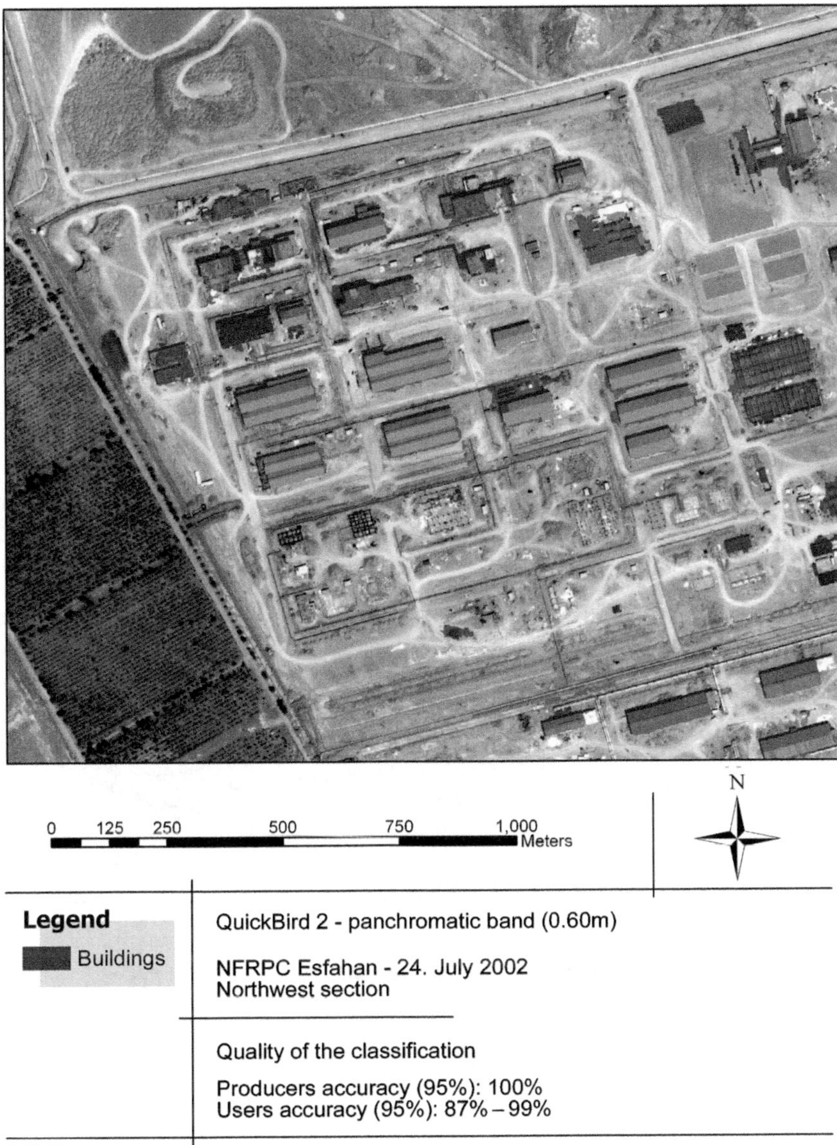

Fig. 9.2 Northwest section of NFRPC Esfahan July 2002 with identified buildings. Original Data/Product © Quickbird/Digital Globe; 2002; Distributed by Eurimage

Furthermore, significant changes over considerable periods of time can be automatically detected and classified. Maps can be generated that show changes in the classes sought, in this case newly constructed buildings and roads (Fig. 9.3).

A large-area search for undeclared industrial facilities is also possible. Low-resolution satellite images covering a much larger area can be used for this purpose.

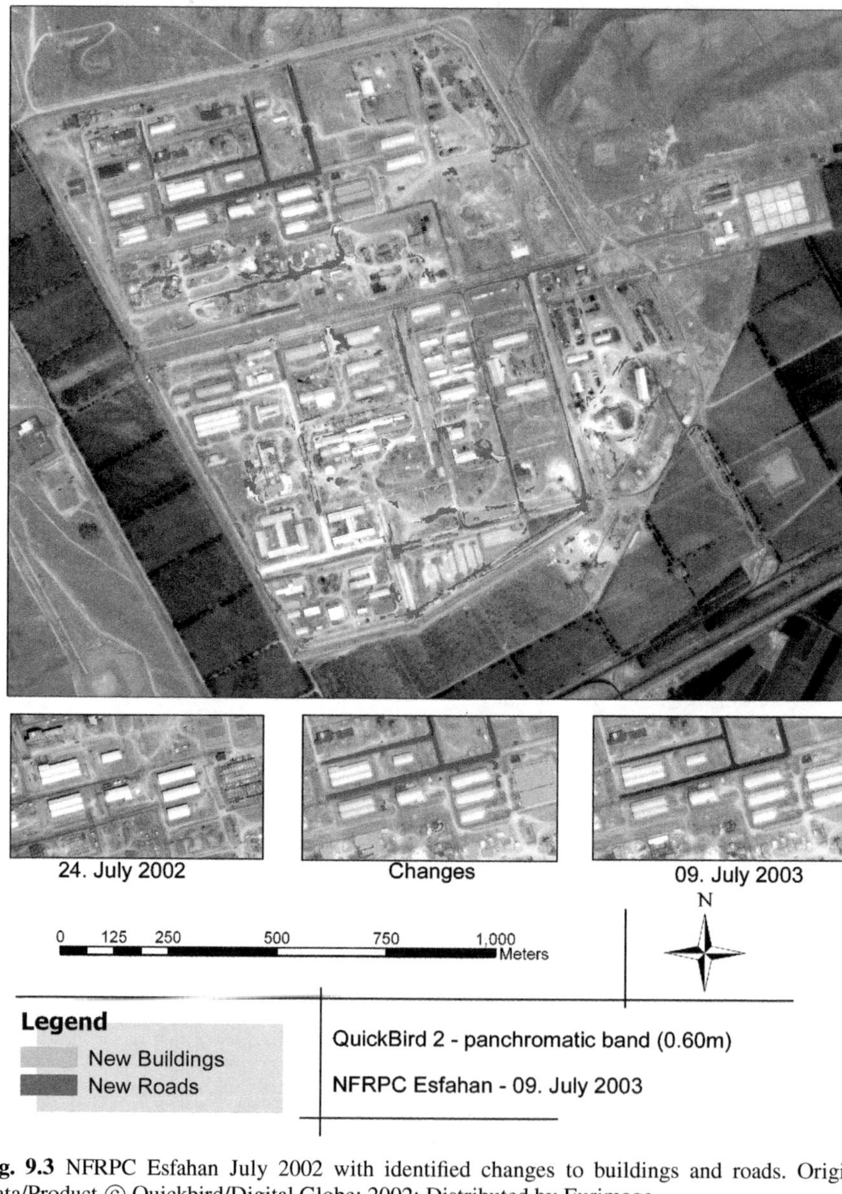

Fig. 9.3 NFRPC Esfahan July 2002 with identified changes to buildings and roads. Original Data/Product © Quickbird/Digital Globe; 2002; Distributed by Eurimage

All existing facilities have been identified in the examinations made. In addition, other industrial facilities have been recognized as potential targets. However, this method also involves misclassifications, i.e. areas are identified as industrial facilities which are not. Since, however, none of the nuclear test facilities were "overlooked" by the model, this approach is also very promising as providing support for the NPT. Figure 9.4 once again shows the region close to Natanz. With the aid of such a map, an image analyst can gain a quick impression and, if required, can also take recourse to a higher-resolution satellite image.

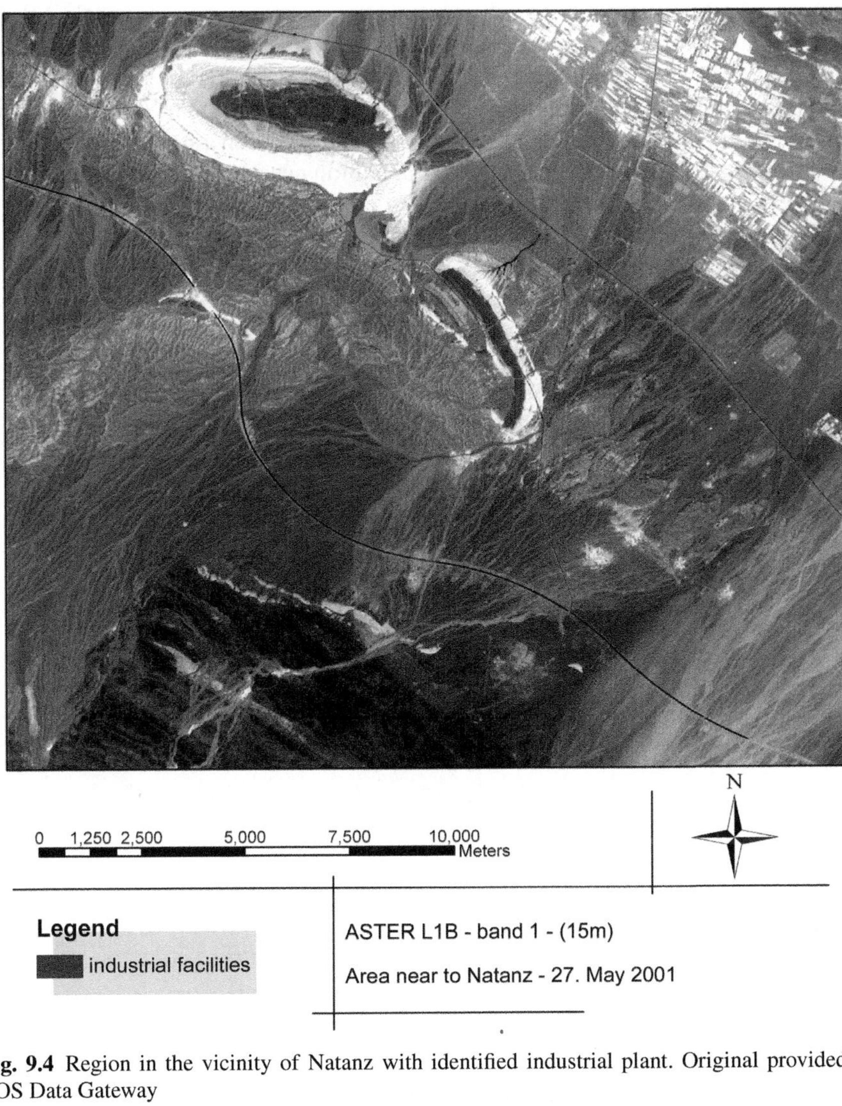

Fig. 9.4 Region in the vicinity of Natanz with identified industrial plant. Original provided by EOS Data Gateway

The methods developed and applied are capable of specifically identifying objects relevant for nuclear verification within the RS data and of detecting and identifying changes over considerable periods of time. As already indicated, computer-controlled verification solely on the basis of satellite images is not yet possible at the present time. The analyses can, however, provide considerable support for the work of a human image analyst.

This thesis has shown that the new approach to remote sensing, object-based image analysis, has a great potential in general and for verification of the NPT in particular. Future work should be concerned with the development of a comprehensive satellite-based verification and interpretation system. This will require characteristic features and reference models for different types of facility in various political and geographical areas. Different types of facility result from the various elements of the nuclear fuel cycle such as uranium mining, enrichment, fuel production, nuclear power plants, reprocessing and storage. The reference models to be developed could then be integrated in an image analysis system and the operations automated. In order to achieve this, the quality of the segmentation needs to be improved in future. To this end, it would be useful to establish a measure of quality for segmentation. The preprocessing procedures should be continuously further developed. There is also a need for research in the field of feature analysis, where SEaTH will have to be expanded to include features of non-normal distribution, and different approaches for feature analysis are also required.

The method of object-based image analysis in the field of remote sensing and the path towards satellite-based verification of the NPT are undoubtedly still just at the start of a process of continuous development. This thesis has shown that both methods are already of great value in handling current issues and, moreover, have extensive potential for future requirements.

"A path is made by treading it"[1]

[1] Chuang Tse (*c*.369–*c*.286 BC), Taoist philosopher.

Appendix A
SEaTH-Analysis of the NFRPC Esfahan

Table A.1 SEaTH-analysis of NFRPC Esfahan, July 2002 – Part 1

Fourth Scalelevel	*Buildings*	*to*	*Vegetation*		
Features	**Separability**	**S1**	**S2**	**Sign**	**m1 > x < m2**
Mean B1	1,74	428,64	169,122	great	428,64
Mean B2	1,73	728,076	172,784	great	728,076
Mean NDVI	1,67	-0,114	0,056	small	0,056
Ratio B2	1,67	0,299	0,37	great	0,299
Mean B3	1,66	607,646	-96,681	great	607,646
Ratio B4	1,55	0,178	0,28	small	0,28
Stddev NDVI	1,43	-0,018	0,022	small	0,022
Fourth Scalelevel	*Buildings*	*to*	*Roads*		
Features	**Separability**	**S1**	**S2**	**Sign**	**m1 > x < m2**
Compactness	1,2	0,027	3,117	small	3,117
Length(Pxl)	1,17	-5,927	171,591	small	171,591
Mean NDVI	1,09	0,065	0,002	small	0,002
Border l.(Pxl)	1,08	-102,085	634,218	small	634,218
Stddev B1	0,89	28,432	-1,052	great	28,432
Ratio B2	0,89	0,313	0,22	great	0,313
Shape index	0,83	-1,189	3,393	small	3,393
Fourth Scalelevel	*Buildings*	*to*	*Background*		
Features	**Separability**	**S1**	**S2**	**Sign**	**m1 > x < m2**
Ratio B2	1,67	0,306	0,269	great	0,306
Mean NDVI	1,65	0,073	0,021	small	0,021
Ratio B4	1,6	0,306	0,262	small	0,262
Mean B1	1,39	547,11	356,718	great	547,11
Mean B2	1,3	941,617	588,358	great	941,617
Area(Pxl)	1,28	-1248,83	3874,93	small	3874,93
Stddev B1	1,27	21,322	-0,949	great	21,322
Fourth Scalelevel	*Vegetation*	*to*	*Roads*		
Features	**Separability**	**S1**	**S2**	**Sign**	**m1 > x < m2**
Stddev NDVI	1,74	0,013	-0,007	great	0,013
Mean NDVI	1,7	0,061	-0,019	great	0,061
Mean B1	1,66	103,301	403,239	small	403,239
Mean B2	1,52	101,482	677,023	small	677,023
Mean B3	1,37	-147,474	563,614	small	563,614
Ratio B4	1,37	0,3	0,219	great	0,3
Ratio B2	1,3	0,332	0,286	small	0,286

Table A.2 SEaTH-analysis of NFRPC Esfahan, July 2002 – Part 2

Fourth Scalelevel	*Vegetation*	*to*	*Background*		
Features	**Separability**	**S1**	**S2**	**Sign**	**m1 > x < m2**
Stddev NDVI	1,85	0,008	–0,002	great	0,008
Mean B1	1,81	1360,16	394,079	small	394,079
Mean B2	1,76	1843,4	659,937	small	659,937
Mean B3	1,72	1213,89	555,567	small	555,567
Mean NDVI	1,71	0,069	0,012	great	0,069
Ratio B3	1,53	0,268	0,243	small	0,243
Ratio B4	1,5	0,296	0,255	great	0,296
Fourth Scalelevel	*Roads*	*to*	*Background*		
Features	**Separability**	**S1**	**S2**	**Sign**	**m1 > x < m2**
Density	1,15	–1,118	1,336	small	1,336
Compactness	0,94	4,257	–0,044	great	4,257
Length/width	0,86	7,609	–3,193	great	7,609
Mean NDVI	0,82	0,082	0,031	small	0,031
Area (Pxl)	0,71	–2698,59	8792,75	small	8792,75
Rectangular Fit	0,62	1,202	0,544	small	0,544
GLCM Ang. 2nd NDVI	0,61	0,001	0	great	0
Second Scalelevel	*Shadows*	*to*	*Buildings*		
Features	**Separability**	**S1**	**S2**	**Sign**	**m1 > x < m2**
Rel. border to brighter	1,22	1	1	great	1
neighbours B3					
Rel. border to brighter	1,22	1	1	great	1
neighbors B1					
Rel. border to brighter	1,21	1	1	great	1
neighbors B2					
Asymmetry	1,21	1	1	great	1
Second Scalelevel	*Shadows*	*to*	*Background*		
Features	**Separability**	**S1**	**S2**	**Sign**	**m1 > x < m2**
Ratio B1	1,78	0,199	0,1047	great	0,199
Ratio B4	1,77	0,589	0,2372	small	0,237
Ratio B2	1,74	0,317	–0,2696	great	0,317
Mean B4	1,72	–591,491	437,478	small	437,478

Table A.3 SEaTH-analysis of NFRPC Esfahan, July 2002 – Part 3

First Scalelevel	*Walls...*	*to*	*Background*		
Features	**Separability**	**S1**	**S2**	**Sign**	**m1 > x < m2**
Stddev B1	1,57	13,468	-7,323	great	13,468
Stddev B2	1,55	25,309	-13,867	great	25,309
Stddev B3	1,5	23,267	-14,762	great	23,267
GLCM B1 Homogeneity (all dir.)	1,49	-0,013	0,029	small	0,029
Mean NDVI	1,49	0,078	0,036	small	0,036
GLCM B3 Correlation (all dir.)	1,49	1,029	0,597	small	0,597
Stddev B4	1,48	25,609	-19,811	great	25,609
First Scalelevel	*Walls...*	*to*	*Roads*		
Features	**Separability**	**S1**	**S2**	**Sign**	**m1 > x < m2**
Stddev B3	1,32	25,964	-12,454	great	25,964
Stddev B4	1,31	28,247	-13,704	great	28,247
GLCM B1 Correlation (all dir.)	1,31	1,264	0,597	small	0,597
Stddev B2	1,31	29,46	-11,478	great	29,46
GLCM B3 Dissimilarity (all dir.)	1,31	44,933	11,099	great	44,933
GLCM B3 Correlation (all dir.)	1,3	1,251	0,57	small	0,57
First Scalelevel	*Walls...*	*to*	*Vegetation*		
Features	**Separability**	**S1**	**S2**	**Sign**	**m1 > x < m2**
Mean NDVI	1,77	-0,004	0,054	small	0,054
Ratio B3	1,72	0,248	0,262	great	0,248
Ratio B4	1,67	0,251	0,285	small	0,285
Ratio B2	1,51	0,288	0,312	great	0,288
Mean B3	1,39	521,89	-11016,2	great	521,89
Stddev NDVI	1,36	-0,003	0,008	small	0,008
Mean B1	1,36	385,055	58,768	great	385,055
First Scalelevel	*Walls...*	*to*	*Buildings*		
Features	**Separability**	**S1**	**S2**	**Sign**	**m1 > x < m2**
Ratio B2	1,79	0,279	0,309	small	0,309
Ratio B4	1,71	0,256	0,29	great	0,256
Mean NDVI	1,64	0,006	0,057	great	0,006
Ratio B3	1,55	0,251	0,264	great	0,251
Ratio B1	1,52	0,163	0,186	small	0,186
GLCM Correl.(all dir.) B1	1,38	1,194	0,606	small	0,606
GLCM Correl.(all dir.) B2	1,31	1,117	0,584	small	0,584

Appendix B
Analysis of Arak, Bandar Abbas, Bushehr and Natanz

Table B.1 Features for the object-based analysis of the Aster data with SEaTH – Part 1

1	Mean Channel 1	34	GLCM Homogeneity (all dir.) Channel 3
2	Stddev Channel 1	35	GLCM Contrast (all dir.) Channel 3
3	Ratio Channel 1	36	GLCM Dissimilarity (all dir.) Channel 3
4	GLCM Homogeneity (all dir.) Channel 1	37	GLCM Entropy (all dir.) Channel 3
5	GLCM Contrast (all dir.) Channel 1	38	GLCM Ang. 2nd moment (all dir.) Channel 3
6	GLCM Dissimilarity (all dir.) Channel 1	39	GLCM Mean (all dir.) Channel 3
7	GLCM Entropy (all dir.) Channel 1	40	GLCM StdDev (all dir.) Channel 3
8	GLCM Ang. 2nd moment (all dir.) Channel 1	41	GLCM Correlation (all dir.) Channel 3
9	GLCM Mean (all dir.) Channel 1	42	GLDV Ang. 2nd moment (all dir.) Channel 3
10	GLCM StdDev (all dir.) Channel 1	43	GLDV Entropy (all dir.) Channel 3
11	GLCM Correlation (all dir.) Channel 1	44	GLDV Mean (all dir.) Channel 3
12	GLDV Ang. 2nd moment (all dir.) Channel 1	45	GLDV Contrast (all dir.) Channel 3
13	GLDV Entropy (all dir.) Channel 1	46	Mean Channel 4
14	GLDV Mean (all dir.) Channel 1	47	Stddev Channel 4
15	GLDV Contrast (all dir.) Channel 1	48	Ratio Channel 4
16	Mean Channel 2	49	GLCM Homogeneity (all dir.) Channel 4
17	Stddev Channel 2	50	GLCM Contrast (all dir.) Channel 4
18	Ratio Channel 2	51	GLCM Dissimilarity (all dir.) Channel 4
19	GLCM Homogeneity (all dir.) Channel 2	52	GLCM Entropy (all dir.) Channel 4
20	GLCM Contrast (all dir.) Channel 2	53	GLCM Ang. 2nd moment (all dir.) Channel 4
21	GLCM Dissimilarity (all dir.) Channel 2	54	GLCM Mean (all dir.) Channel 4
22	GLCM Entropy (all dir.) Channel 2	55	GLCM StdDev (all dir.) Channel 4
23	GLCM Ang. 2nd moment (all dir.) Channel 2	56	GLCM Correlation (all dir.) Channel 4
24	GLCM Mean (all dir.) Channel 2	57	GLDV Ang. 2nd moment (all dir.) Channel 4
25	GLCM StdDev (all dir.) Channel 2	58	GLDV Entropy (all dir.) Channel 4
26	GLCM Correlation (all dir.) Channel 2	59	GLDV Mean (all dir.) Channel 4
27	GLDV Ang. 2nd moment (all dir.) Channel 2	60	GLDV Contrast (all dir.) Channel 4
28	GLDV Entropy (all dir.) Channel 2	61	Mean Channel 5
29	GLDV Mean (all dir.) Channel 2	62	Stddev Channel 5
30	GLDV Contrast (all dir.) Channel 2	63	Ratio Channel 5
31	Mean Channel 3	64	GLCM Homogeneity (all dir.) Channel 5
32	Stddev Channel 3	65	GLCM Contrast (all dir.) Channel 5
33	Ratio Channel 3	66	GLCM Dissimilarity (all dir.) Channel 5

Table B.1 (continued)

67	GLCM Entropy (all dir.) Channel 5	82	GLCM Entropy (all dir.) Channel 6
68	GLCM Ang. 2nd moment (all dir.) Channel 5	83	GLCM Ang. 2nd moment (all dir.) Channel 6
69	GLCM Mean (all dir.) Channel 5	84	GLCM Mean (all dir.) Channel 6
70	GLCM StdDev (all dir.) Channel 5	85	GLCM StdDev (all dir.) Channel 6
71	GLCM Correlation (all dir.) Channel 5	86	GLCM Correlation (all dir.) Channel 6
72	GLDV Ang. 2nd moment (all dir.) Channel 5	87	GLDV Ang. 2nd moment (all dir.) Channel 6
73	GLDV Entropy (all dir.) Channel 5	88	GLDV Entropy (all dir.) Channel 6
74	GLDV Mean (all dir.) Channel 5	89	GLDV Mean (all dir.) Channel 6
75	GLDV Contrast (all dir.) Channel 5	90	GLDV Contrast (all dir.) Channel 6
76	Mean Channel 6	91	Mean Channel 7
77	Stddev Channel 6	92	Stddev Channel 7
78	Ratio Channel 6	93	Ratio Channel 7
79	GLCM Homogeneity (all dir.) Channel 6	94	GLCM Homogeneity (all dir.) Channel 7
80	GLCM Contrast (all dir.) Channel 6	95	GLCM Contrast (all dir.) Channel 7
81	GLCM Dissimilarity (all dir.) Channel 6	96	GLCM Dissimilarity (all dir.) Channel 7

Table B.2 Features for the object-based analysis of the Aster data with SEaTH - Part 2

97	GLCM Entropy (all dir.) Channel 7	126	GLCM Dissimilarity (all dir.) Channel 9
98	GLCM Ang. 2nd moment (all dir.) Channel 7	127	GLCM Entropy (all dir.) Channel 9
99	GLCM Mean (all dir.) Channel 7	128	GLCM Ang. 2nd moment (all dir.) Channel 9
100	GLCM StdDev (all dir.) Channel 7	129	GLCM Mean (all dir.) Channel 9
101	GLCM Correlation (all dir.) Channel 7	130	GLCM StdDev (all dir.) Channel 9
102	GLDV Ang. 2nd moment (all dir.) Channel 7	131	GLCM Correlation (all dir.) Channel 9
103	GLDV Entropy (all dir.) Channel 7	132	GLDV Ang. 2nd moment (all dir.) Channel 9
104	GLDV Mean (all dir.) Channel 7	133	GLDV Entropy (all dir.) Channel 9
105	GLDV Contrast (all dir.) Channel 7	134	GLDV Mean (all dir.) Channel 9
106	Mean Channel 8	135	GLDV Contrast (all dir.) Channel 9
107	Stddev Channel 8	136	Mean Slope
108	Ratio Channel 8	137	Stddev Slope
109	GLCM Homogeneity (all dir.) Channel 8	138	Ratio Slope
110	GLCM Contrast (all dir.) Channel 8	139	GLCM Homogeneity (all dir.) Slope
111	GLCM Dissimilarity (all dir.) Channel 8	140	GLCM Contrast (all dir.) Slope
112	GLCM Entropy (all dir.) Channel 8	141	GLCM Dissimilarity (all dir.) Slope
113	GLCM Ang. 2nd moment (all dir.) Channel 8	142	GLCM Entropy (all dir.) Slope
114	GLCM Mean (all dir.) Channel 8	143	GLCM Ang. 2nd moment (all dir.) Slope
115	GLCM StdDev (all dir.) Channel 8	144	GLCM Mean (all dir.) Slope
116	GLCM Correlation (all dir.) Channel 8	145	GLCM StdDev (all dir.) Slope
117	GLDV Ang. 2nd moment (all dir.) Channel 8	146	GLCM Correlation (all dir.) Slope
118	GLDV Entropy (all dir.) Channel 8	147	GLDV Ang. 2nd moment (all dir.) Slope
119	GLDV Mean (all dir.) Channel 8	148	GLDV Entropy (all dir.) Slope
120	GLDV Contrast (all dir.) Channel 8	149	GLDV Mean (all dir.) Slope
121	Mean Channel 9	150	GLDV Contrast (all dir.) Slope
122	Stddev Channel 9	151	Mean NDVI
123	Ratio Channel 9	152	Stddev NDVI
124	GLCM Homogeneity (all dir.) Channel 9	153	Ratio NDVI
125	GLCM Contrast (all dir.) Channel 9	154	GLCM Homogeneity (all dir.) NDVI

Table B.2 (continued)

155	GLCM Contrast (all dir.) NDVI	174	Rectangular Fit
156	GLCM Dissimilarity (all dir.) NDVI	175	Border length (m)
157	GLCM Entropy (all dir.) NDVI	176	Shape index.) NDVI
158	GLCM Ang. 2nd moment (all dir.)NDVI	177	Density
159	GLCM Mean (all dir.) NDVI	178	Main direction
160	GLCM StdDev (all dir.) NDVI	179	Asymmetry
161	GLCM Correlation (all dir.) NDVI	180	GLCM Homogeneity (all dir.)
162	GLDV Ang. 2nd moment (all dir.) NDVI	181	GLCM Contrast (all dir.)
163	GLDV Entropy (all dir.) NDVI	182	GLCM Dissimilarity (all dir.)
164	GLDV Mean (all dir.) NDVI	183	GLCM Entropy (all dir.)
165	GLDV Contrast (all dir.) NDVI	184	GLCM Ang. 2nd moment (all dir.)
166	Brightness	185	GLCM Mean (all dir.)
167	Max.Diff.	186	GLCM StdDev (all dir.)
168	Area (m^2)	187	GLCM Correlation (all dir.)
169	Length (m)	188	GLDV Ang. 2nd moment (all dir.)
170	Width (m)	189	GLDV Entropy (all dir.)
171	Length/width	190	GLDV Mean (all dir.)
172	Compactness	191	GLDV Contrast (all dir.)
173	Elliptic Fit		

Table B.3 SEaTH-analysis of Arak, Bandar Abbas, Bushehr and Natanz – Part 1

Industrial Site to		*Soil*			
Features	**Separability**	**S1**	**S2**	**Sign**	**m1 > x < m2**
GLCM Ang. 2nd (all dir.) Channel 1	1.17	−0.002	0.004	small	0.004
GLCM Ang. 2nd (all dir.) Channel 2	1.13	−0.001	0.003	small	0.003
Stddev Channel 3	1.06	9.877	−2.286	great	9.877
GLCM Ang. 2nd (all dir.) Channel 6	1.05	−0.001	0.006	small	0.006
GLCM Ang. 2nd (all dir.) Channel 7	1.05	−0.001	0.007	small	0.007
Industrial Site to		*Mountains*			
Features	**Separability**	**S1**	**S2**	**Sign**	**m1 > x < m2**
GLCM Entropy (all dir.) Slope	1.52	14.407	7.917	small	7.917
GLDV Entropy (all dir.) Channel 7	1.4	2.44	4.008	great	2.44
Mean Slope	1.32	−1.213	8.893	small	8.893
GLDV Entropy (all dir.) Channel 6	1.3	2.534	3.933	great	2.534
GLDV Entropy (all dir.) Channel 8	1.26	2.471	4.289	great	2.471
Industrial Site to		*Settlements*			
Features	**Separability**	**S1**	**S2**	**Sign**	**m1 > x < m2**
GLCM Ang. 2nd (a.d.) Channel 1	1.12	0.001	0	great	0.001
GLCM Homogeneity(a.d.)Chan. 2	1.04	0.05	0.013	great	0.05
GLCM Ang. 2nd (a.d.) Channel 2	1.03	0.001	0	great	0.001
GLCM Homogeneity (a.d.)Chan. 1	1	0.051	0.015	great	0.051
GLCM StdDev (a.d.) Channel 3	0.95	44.327	40.25	small	40.25
Industrial Site to		*Vegetation*			
Features	**Separability**	**S1**	**S2**	**Sign**	**m1 > x < m2**
Ratio Channel 3	1.86	−1.872	0.123	small	0.123
Mean NDVI	1.7	1.352	−0.017	small	−0.017
Mean Channel 3	1.67	−2594.35	102.733	small	102.733
Stddev NDVI	1.4	−0.054	0.079	small	0.079
GLCM StdDev (a.d.) Channel 1	1.29	44.545	41.732	small	41.732
Industrial Site to		*Water*			
Features	**Separability**	**S1**	**S2**	**Sign**	**m1 > x < m2**
Mean Channel 3	2	50.53	−71.812	great	50.53
GLCM Homogeneity(a.d.)Chan. 3	2	1.314	0.155	small	0.155
GLCM Entropy (a.d.) Channel 3	2	4.426	−4.864	great	4.426
Mean Channel 4	2	37.067	−36.393	great	37.067
Ratio Channel 4	2	0.078	0.494	great	0.078

Table B.4 SEaTH-analysis of Arak, Bandar Abbas, Bushehr and Natanz – Part 2

Soil to		Mountains			
Features	**Separability**	**S1**	**S2**	**Sign**	**m1 > x < m2**
Stddev Slope	1.32	45.402	8.853	small	8.853
GLCM Entropy (all dir.) Slope	1.3	11.006	7.972	small	7.972
Mean Slope	1.28	-6.46	9.533	small	9.533
GLCM StdDev (all dir.) Slope	1.21	46.121	38.681	small	38.681
GLCM Ang. 2nd (all dir.) Slope	0.96	0.004	-0.003	great	0.004
Soil to		*Settlements*			
Features	**Separability**	**S1**	**S2**	**Sign**	**m1 > x < m2**
GLCM Ang. 2nd (a.d.) Channel 1	1.73	0.001	0	great	0.001
GLCM Ang. 2nd (a.d.) Channel 2	1.68	0.001	0	great	0.001
GLDV Ang. 2nd (a.d.) Channel 2	1.46	0.045	0.018	great	0.045
Mean Channel 5	1.45	88.7	57.249	great	88.7
GLCM Ang. 2nd (a.d.) Channel 8	1.42	0.004	0.001	great	0.004
Soil to		*Vegetation*			
Features	**Separability**	**S1**	**S2**	**Sign**	**m1 > x < m2**
Ratio Channel 3	1.88	2.158	0.123	small	0.123
Stddev NDVI	1.67	-0.064	0.064	small	0.064
Mean NDVI	1.55	0.802	0.008	small	0.008
Brightness	1.33	98.562	73.201	great	98.562
GLCM StdDev (all dir.)	1.26	33.386	86.832	great	33.386
Soil to		*Water*			
Features	**Separability**	**S1**	**S2**	**Sign**	**m1 > x < m2**
Mean Channel 3	2	47.756	-23.536	great	47.756
Mean Channel 4	2	31.66	-9.811	great	31.66
Ratio Channel 4	2	0.082	0.295	great	0.082
Ratio Channel 6	2	0.074	-0.155	great	0.074
Mean Channel 5	2	32.408	-9.896	great	32.408
Mountains to		*Settlements*			
Features	**Separability**	**S1**	**S2**	**Sign**	**m1 > x < m2**
Stddev Slope	1.78	7.745	-4.247	great	7.745
GLCM Entropy (a.d.) Slope	1.75	7.804	18.495	great	7.804
GLCM Homogeneity(a.d.)Channel 9	1.73	0.093	0.019	great	0.093
GLCM Ang. 2nd (a.d.) Channel 1	1.73	0.001	0	great	0.001
GLCM Ang. 2nd (a.d.) Channel 2	1.69	0.001	0	great	0.001

Table B.5 SEaTH-analysis of Arak, Bandar Abbas, Bushehr and Natanz – Part 3

	Mountains to		*Vegetation*		
Features	**Separability**	**S1**	**S2**	**Sign**	**m1 > x < m2**
Stddev NDVI	1.71	−0.047	0.058	small	0.058
GLCM StdDev (all dir.)	1.42	33.792	159.092	great	33.792
Stddev Slope	1.37	8.609	−196.954	great	8.609
Mean Slope	1.35	8.771	−1.273	great	8.771
GLCM Entropy (all dir.) Slope	1.33	7.96	12.364	great	7.96
	Mountains to		*Water*		
Features	**Separability**	**S1**	**S2**	**Sign**	**m1 > x < m2**
Ratio Channel 6	2	0.07	−0.016	great	0.07
Ratio Channel 4	2	0.07	−0.012	great	0.07
GLCM Homogeneity(a.d.)Chan. 3	2	0.452	0.176	small	0.176
Mean Channel 9	2	39.126	−19.519	great	39.126
Mean Channel 3	2	41.315	−1.501	great	41.315
	Settlements to		*Vegetation*		
Features	**Separability**	**S1**	**S2**	**Sign**	**m1 > x < m2**
Ratio Channel 3	1.98	0.08	0.115	small	0.115
Mean NDVI	1.97	−0.382	−0.071	small	−0.071
Mean Channel 3	1.88	1004.67	99.733	small	99.733
GLCM Entropy (all dir.)	1.77	8.821	6.67	great	8.821
GLCM Ang. 2nd (all dir.)	1.61	0	0	small	0
	Settlements to		*Water*		
Features	**Separability**	**S1**	**S2**	**Sign**	**m1 > x < m2**
Ratio Channel 1	2	0.118	0.211	small	0.211
GLCM Entropy (a.d.) Channel 1	2	6.636	14.012	great	6.636
GLCM Entropy (a.d.) Channel 2	2	6.554	12.997	great	6.554
GLDV Entropy (a.d.) Channel 2	2	2.688	25.361	great	2.688
Mean Channel 3	2	47.849	−55.131	great	47.849

Bibliography

Abmayer, W. (1994). *Einführung in die digitale Bildverarbeitung*. Teubner, Stuttgart.

Aiazzi, B., Alparone, L., Baronti, S., and Garzelli, A. (2002). Context-driven fusion of high spatial and spectral resolution data based on oversampling multiresolution analysis. *IEEE Transactions on Geoscience and Remote Sensing*, **40**(10), 2300–2312.

Albertz, J. (2001). *Einführung in die Fernerkundung. Grundlagen der Interpretation von Luft- und Satellitenbildern*. Wissenschaftliche Buchgesellschaft, Darmstalt.

Avenhaus, R., Kyriakopoulos, N., Richard, M., and Stein, G. (2006). *Verifying Treaty Compliance – Limiting Weapons of Mass Destruction and Monitoring Kyoto Protocol Provisions*. Springer, New York.

Baatz, M. and Schäpe, A. (1999). Object-oriented and multi-scale image analysis in semantic networks. *Proceedings of the 2nd International Symposium on Operationalization of Remote Sensing, 16–20 August, Enschede, The Netherlands. Enschede, CD-Rom.*

Baatz, M. and Schäpe, A. (2000). Multiresolution segmentation – an optimization approach for high quality multi-scale image segmentation. In: J. Strobl, and T. Blaschke (Eds.): *Angewandte Geographische Informationsverarbeitung XII*. Beiträge zum AGIT-Symposium Salzburg. Heidelberg, New York.

Baatz, M., Heyen, M., and Hofman, P. (2004). *eCognition – Object oriented image analysis*. User Guide. Definiens AG. Abrufbar unter: http://www.definiens-imaging.com.

Banko, G. (1998). A review of assessing the accuracy of classifications of remotely sensed data and methods including remote sensing data in forest inventory. *Interim Report IR-98-081. IIASA, Laxanburg, Austria.*

Baute, J. G. (2006). A concrete experience: The Iraq Case. In: *Verifying Treaty Compliance – Limiting Weapons of Mass Destruction and Monitoring Kyoto Protocol Provisions*, R. Avenhaus, N. Kyriakopoulos, M. Richard and G. Stein (Ed.), Springer, New York.

Benz, U., Hofmann, P., Willhauck, G., Lingenfelder, I., and Heynen, M. (2004). Multi-resolution, object-oriented fuzzy analysis of remote sensing data for gis-ready information. *ISPRS Journal of Photogrammetry & Remote Sensing* **58** (2004), 239–258.

Berk, A., Bernstein, L., and Robertson, D. C. (1989). Modtran: A moderate resolution model for lowtran 7. afgl-tr-89-0122. *Air Force Geophysics Laboratory, Hanscom AFB MA 01731.*

Bhattacharyya, A. (1943). On a measure of divergence between two statistical populations defined by their probability distributions. *Bulletin of the Calcutta Mathematical Society*, **(35)**, 99–110.

Bitelli, G., Camassi, R., Gusella, R., and Mongol, A. (2004). Image change detection on urban area: The earthquake case. *Proceedings of the ISRPS 2004 Annual Conference, Istanbul, Turkey*. July 19–23.

Blaschke, T. (2000). *Objektextraktion und regelbasierte Klassifikation von Fernerkundungsdaten: Neue Möglichkeiten für GIS-Anwender und Planer*. In: M. Schrenk, (Ed.): CORP'2000. Technical University of Vienna.

Blaschke, T. (2004). Towards a framework for change detection based on image objects. *1st GGRS, Göttingen*, 7–8 October.

Blaschke, T. (2005). A framework for change detection based on image objects. In: *Göttinger Geographische Abhandlungen*, S., Erasmi, B., Cyffka, M. Kappas, (Eds.) **113**, 1–9. Goltze, Goettingen.

Bruzzone, L. and Prieto, D. F. (1999). A bayesin appraoch to automatic change detection. *Proceedings of the IEEE 1999 International Geoscience and Remote Sensing Symposium, Hamburg.*

Byrne, G. F., Crapper, P. F., and Mayo, K. K. (1980). Monitoring land-cover changes by principal component analysis of multitemporal landsat data. *Remote Sensing of Environment*, **10**, 175–184.

Cabral, P., Gilg, J., and Painho, M. (2005). Monitoring urban growth using remote sensing, gis and spatial metrics, remote sensing and modeling of ecosystems for sustainability. *Proceedings of SPIE – Optics & Photonics, San Diego, USA, 29 July–4 August.*

Campbell, J. B. (1996). *Introduction to Remote Sensing*, 2nd Edition, Taylor & Francis, New York.

Canty, M. J. (1999). *Fernerkundung mit Neuronalen Netzen*. Expert-Verlag, Germany.

Canty, M. J. and Schlittenhardt, J. (2001). Satellite data used to locate site of 1998 indian nuclear test. *Eos, Transactions, American Geophysical Union*, **82**(3), 25–29.

Canty, M. J., Nielsen, A. A., and Schmidt, M. (2004). Automatic radiometric normalization of multispectral imagery. *Remote Sensing of Environment*, **91**(3,4), 441–451.

Carlson, J. (2006). Experience and challenges in WMD treaty verification: A comparative view. In: *Verifying Treaty Compliance – Limiting Weapons of Mass Destruction and Monitoring Kyoto Protocol Provisions*, R. Avenhaus, N. Kyriakopoulos, M. Richard and G. Stein (Eds.), Springer, New York.

Centeno, J. S. (1998). Integrierte Verarbeitung von Satellitenbild- und gerasteter Karteninformation. In: *Digitale Bildverarbeitung*. H.-P. Bähr and T. Vögtle (Eds.): Anwendung in Photogrammetrie, Kartographie und Fernerkundung, Heidelberg.

Chandra, M., Moreira, A., and Keydel, W. (2005). Amper: Network on applied multiparameter environmental remote sensing. An eu sponsored resaerch and training network. *Proceedings of the IGARSS 2005 Symposium. Seoul, Korea.*

Chavez, P. and MacKinnon, D. J. (1994). Automatic detection of vegetation changes in the southwestern united states using remotely sensed images. *Photogrammetric Engineering and Remote Sensing*, **60**(5), 571–583.

Chen, Y., Fung, T., Lin, W., and Wang, J. (2005). An image fusion method based on object-oriented image classification. *Proceedings of the IGARSS 2005 Symposium. Seoul, Korea*, 25–29 July.

Chunyang, H., Li, J., Zhang, J., Pan, Y., and Chen, Y. H. (2005). Dynamic monitor on urban expansion based on a object-oriented approach. *Proceedings of the IGARSS 2005 Symposium. Seoul, Korea*, 25–29 July.

Clayton, D. G. (1971). The gram-schmidt othogonalization. *Applied Statistics*, **20**, 335–338.

Congalton, R. G. (1991). A review of assessing the accuracy of classifications of remotely sensed data. *Remote Sensing Environment*, **37**, 35–46.

Cooley (2006). Nuclear nonproliferation treaty NPT. in: *Verifying Treaty Compliance – Limiting Weapons of Mass Destruction and Monitoring Kyoto Protocol Provisions*, R. Avenhaus, N. Kyriakopoulos, M. Richard and G. Stein (Eds.), Springer, New York.

Crase, B. and Hempel, C. (2005). Object based land cover mapping for groote eylandt: a tool for reconnaissance and land based surveys. *Proceedings of NARGIS 2005 – APPLICATIONS IN TROPICAL SPATIAL SCIENCE. 4th–7th July Charles Darwin University, Darwin, NT, Australia.*

Dai, X. and Khorram, S. (1997). Development of a feature-based approach to automated image registration for multitemporal and multisensor remotely sensed imagery. *Proceedings of the IEEE 1997 Internatinal Geoscience and Remote Sensing Symposium, Singapore.*

Dai, X., Khorram, S., and Cheshire, H. (1996). Automated image registration for change detection from landsat thematic mapper imagery. *Proceedings of the IEEE 1996 Internatinal Geoscience and Remote Sensing Symposium, Lincoln, USA.*

Du, Y., Teillet, P., and Cihlar, J. (2002). Radiometric normalization of multitemporal high-resolution images with quality control for land cover change detection. *Remote Sensing Environment*, **82**, 123–134.

Eklundh, L. and Singh, A. (1993). A comparative analysis of standardised and unstandardised principal components analysis in remote sensing. *International Journal of Remote Sensing*, **14**(7), 1359–1370.

Farebrother, R. W. (1974). The gram-schmidt regression. *Applied Statistics*, **23**, 470–476.

Feitosa, R., Costa, G., and Cazes, T. (2006). A genetic approach for the automatic adaptation of segmentation parameters. *Proceedings of the AGIT – Obia, Salzburg*, 4–5 July.

Fischer, W. and Stein, G. (1998). Das gestärkte safeguards-system der iaeo. *Energiewirtschaftliche Tagesfragen*, **48**(8), 524–530.

Frauman, E. and Wolf, E. (2005). Segmentation of very high spatial resolution satellite images in urban areas for segments-based classification. *Proceedings of the ISPRS WG VII/1 Human Settlements and Impact Analysis 3rd International Symposium Remote Sensing and Data Fusion Over Urban Areas (URBAN 2005) and 5th International Symposium Remote Sensing of Urban Areas (URS 2005). Tempe, AZ, USA.*

Freund, J. (2003). Mathematical statistics with Applications. *Prentice-Hall International.*

Förstner, W. and Gülch, E. (1987). A fast operator for detection and precise location of distinct points, corners and centers of circular features. *ISPRS Intercommission Workshop, Interlaken.*

Förstner, W., Liedtke, C.-E., and Bückner J. (Eds.) (1999). SMATI 99. *Proceedings of the Semantic Modelling for the Acquisition of Topographic Information from Images and Maps.* Munich, Germany.

Fukunaga, K. (1990). *Introduction to Statistical Pattern Recognition*, 2nd Edition. Academic Press Inc, Burlington, MA, USA.

Fung, T. (1990). An assessment of tm imagery for land-cover change detection. *IEEE Transactions on Geoscience and Remote Sensing*, **28**(4), 681–684.

Fung, T. and LeDrew, E. (1987). Application of principal components analysis to change detection. *Photogrammetric Engineering & Remote Sensing*, **53**(12), 1649–1658.

Fung, T. and LeDrew, E. (1988). The determination of optimal threshold levels for change detection using various accuracy indices. *Photogrammetric Engineering & Remote Sensing*, **54**(10), 1449–1454.

Furby, S. and Campell, N. (2001). Calibrating images from different dates to like-value counts. *Remote Sensing Environment*, **82**, 123–134.

Gong, P. (1993). Change detection using principal components analysis and fuzzy set theory. *Canadian Journal of Remote Sensing*, **19**(1), 22–29.

Greiwe, A. (2004). Feature-based fusion of multisensor data – inclusion of hyperspectral data into classification of high resolution orthophotos. *1st GGRS, Gvttingen*, 7–8 October.

Grenzdoerfer, G. (2005). Land use change in rostock, germany since the reunification – a combined approach with satellite data and high resolution aerial images. *Proceedings of the ISPRS WG VII/1 Human Settlements and Impact Analysis, 3rd International Symposium Remote Sensing and Data Fusion Over Urban Areas (URBAN 2005) and 5th International Symposium Remote Sensing of Urban Areas (URS 2005). Tempe, AZ, USA*, 14–16 March.

Growe, S. and Tönjes, R. (1997). A knowledge based approach to automatic image registration. *Proceedings of the International Conference on Image Processing 97 (ICIP 97), Santa Barbara, USA.*

Hall, F., Strebel, D., Nickeson, J., and Goetz, S. (1991). Radiometric rectification: toward a common radiometric response among multidate, multisensor images. *Remote Sensing Environment*, **82**, 123–134.

Haralick, R. and Shapiro, L. (1992). *Computer and Robot Vision.* Volume 1. Addison-Wesley, Reading.

Haralick, R. and Shapiro, L. (1993). *Computer and Robot Vision.* Volume 2. Addison-Wesley, Reading.

Heremans, R., Willekens, A., Borghys, D., Verbeeck, B., Valckenborgh, J., and Perneel, C. (2005). Automatic detection of flooded areas on envisat/asar images using an object-oriented classification technique and an active contour algorithm. *Proceedings of the 31th International Symposium on Remote Sensing of the Environment.* June 20–24, *St. Petersburg, Russian Federation.*

Hofmann, T. and Buhmann, J. M. (1997). Pairwise data clustering by deterministic annealung. *IEEE Transactions on Pattern Analysis and Machine Intelligence*, **PAMI-19**(1), 1–4.

Hofmann, T., Puzicha, J., and Buhmann, J. M. (1996). Unsupervised segmentation of textured images by pairwise data clustering. *Proceedings of the IEEE Conference on Image Processing, 1996, Lausanne, Switzerland*, **III**, 137–140.

Hudson, W. D. and Ramm, C. W. (1987). Correct formulation of the kappa coefficient of agreement. *Photogrammetric Engineering and Remote Sensing*, **53**(4), 421–422.

IAEA/Hg. (1968). The agency's safeguards system. *Information Circular 66*, Rev. 2, Wien.

IAEA/Hg. (1970). Treaty on the non-proliferation of Nuclear Weapons. Information Circular 140, Wien.

IAEA/Hg. (1972). The structure and content of agreements between the agency and states required in connection with the treaty on the non-proliferation of Nuclear Weapons. *Information Circular153*, Wien.

IAEA/Hg. (1998). Model protocol additional to the agreement(s) between state(s) and the international atomic energy agency for the application of safeguards. *Information Circular 540 (Corrected)*, Wien.

Ingebritsen, S. E. and Lyon, R. J. P. (1985). Principal components analysis of multitemporal image pairs. *International Journal of Remote Sensing*, **6**(5), 687–696.

Jasani, B. (2004). Identification of key features of nuclear reactors for interpretation of imageries from remote sensing satellites. *Journal of Nuclear Materials Management*, **32**(3), 28–37.

Jasani, B. (2006). Civil reconnaissance satellites – opportunities and challenges, In: *Verifying Treaty Compliance – Limiting Weapons of Mass Destruction and Monitoring Kyoto Protocol Provisions*, R. Avenhaus, N. Kyriakopoulos, M. Richard and G. Stein (Eds.). Springer, New York.

Jensen, J. R. (2004). *Introductory Digital Image Processing – A Remote Sensing Perspective*, 3rd Edition, Prentice Hall, Upper Saddle River, NJ, USA.

Jiaju, L. (1988). Development of principal component analysis applied to multitemporal landsat tm data. *International Journal of Remote Sensing*, **9**(12), 1895–1907.

Kneizys, F., Shettel, E., Abreu, L., Chetwynd, J., Anderson, G., Gallery, W., Selby, J., and Clough, S. A. (1988). Users guide to lowtran 7. afgltr-88-0177. *Air Force Geophysics Laboratory, NTIS No. AD-A206773.*

Koch, B., Jochum, M., Ivitis, E., and Dees, M. (2003). Pixelbasierte Klassifizierung im Vergleich and zur Ergänzung zum objektbasierten Verfahren. *Photogrammetrie Fernerkundung Geoinformation*, **3**, 195–204.

Kosugi, Y. and Kosaka, N. (2005). Development of agricultural gis on shonai area in northeast japan using satellite data. *Proceedings of the 26th Asian Conference on Remote Sensing. Hanoi, Vietnam*, 7–11 November.

Kouchi, K. and Yamazaki, F. (2005). Damage detection based on object-based segmentation and classification from high-resolution satellite images for the 2003 boumerdes, algeria earthquake. *Proceedings of the 26th Asian Conference on Remote Sensing. Hanoi, Vietnam*, 7–11 November.

Kraus, K. (1990). *Fernerkundung Band 2, Auswertung photographischer und digitaler Bilder.* F. Bonn: Ferd. Dümmlers Verlag.

Kraus, K. and Schneider, W. (2001). *Fernerkundung. Physikalische Grundlagen. Band 1.* Dümmler Verlag, Bonn.

Kronfeld, U., Baus, W., Ebbesen, B., and Jathe M. (Eds.), (1993). *Abrüstung, Konversion, Verifkation, Technikfolgen, Ökologische Sicherheit. Naturwissenschaftliche Forschungsprojekte an deutschen Hochschulen.* Kieler Schriften zur Friedenswirtschaft, Band 2, LIT, Hamburg.

Kunz, D., Vögtle, T., and Schilling, K.-J. (1998). Integrierte verarbeitung von satellitenbild und vektorieller karteninformation. in: H.-P. Bähr und T. Vögtle (Ed.): *Digitale Bildverarbeitung. Anwendung in Photogrammetrie, Kartographie und Fernerkundung*, Heidelberg, New York.

Laben, C. A. and Brower, B. V. (2000). *Process for enhancing the spatial Resolution of multispectral Imagery using Pan-sharpening*. United States Patent.

Laliberte, A., Rango, A. and Fredrickson, E. (2005). Classification of arid rangelands using an object-oriented and multi-scale approach with quickbird imagery. *Proceedings of the ASRPS 2005 Annual Conference, Baltimore, MD, USA*, 7–11 March.

Langanke, T., Blaschke, T., and Lang, S. (2004). An object-based gis / remote sensing approach supporting monitoring tasks in european-wide nature conservation. *Proceedings of the Mediterranean conference on Earth Observation. First Mediterranean Conference on Earth Observation (Remote Sensing)*, 21–23, April Belgrade, 245–252.

Lee, A. J., Carender, N. H., Knowlton, D. J., Bell, D. M., and Bryan, J. K. (1993). Autonomous registration of landsat, spot, and digital map imagery. *Proceedings of the SPIE 1944: Integrating Photogrammetric Techniques with Scene Analysis and Machine Vision, April, Orlando, USA*, 68–79.

Lehner, M. (1986). Triple stereoscopic imagery simulation and digital image correlation for meoss project. *Proceedings of the ISPRS Commision I Symposium, Stuttgart*, 477–484.

Lehner, M. and Gill, R. (1989). *Photogrammetric Adjustment of Triple Stereoscopic Imagery of an Airbone CCD Scanner*. Proceedings of the Optical 3-D Measurment Techniques, Wichman Verlag, Karlsruhe.

Lehner, M. and Gill, R. (1992). Semi-automatic derivation of digital elevation models from stereoscopic 3-line scanner data. *Proceedings of Satellite Symposia 1 & 2: Navigation & Mobile Communications, and Image Processing, GIS & Space-assisted Mapping, International Space Year Conference, Munich, Germany*.

LeMoigne, J. (1998). First evaluation of automatic image registration methods. *Proceedings of the IEEE 1998 International Geoscience and Remote Sensing Symposium, Seattle, USA*, 6–10 July.

Li, H., Manjunath, B. S., and Mitra, S. K. (1995). A contour-based approach to multisensor image registration. *IEEE Transactions on Image Processing*, **4**(3), 320–334.

Lillesand, T. M., Kiefer, R. W., and Chipman, J. W. (2003). *Remote Sensing and Image Interpretation*. John Wiley & Sons, Hoboken.

Liu, Z., Wang, J., and Liu, W. (2005). Building extraction from high resolution imagery based on multi-scale object oriented classification and probabilistic hough transform. *Proceedings of the IGARSS 2005 Symposium. Seoul, Korea*, 25–29 July.

Macleod, R. D. and Congalton, R. G. (1998). A quantitative comparison of change detection algorithms for monitoring eelgrass from remotely sensed data. *Photogrammetric Engineering & Remote Sensing*, **64**(3), 207–216.

Mallat, S. G. (1989). A theory for multiresolution signal decomposition: The wavelet representation. *IEEE Transactions on Pattern Analysis and Machine Intelligence*, **11**(7), 674–693.

Mas, J.-F. (1999). Monitoring land-cover changes – a comparison of change detection techniques. *International Journal of Remote Sensing*, **20**(1), 139–152.

Mather, P. M. (2004). *Computer Processing of Remotely Sensed Images. An Introduction*. John Wiley & Sons, Chichester.

Metternicht, G. (1999). Change detection assessment using fuzzy sets and remotely sensed data: An application of topographic map revison. *ISPRS Journal of Photogrammetry & Remote Sensing*, **54**(4), 221–233.

Moeller, M. (2005). Remote sensing for the monitoring of urban growth patterns. *Proceedings of the ISPRS WG VII/1 Human Settlements and Impact Analysis 3rd International Symposium Remote Sensing and Data Fusion Over Urban Areas (URBAN 2005) and 5th International Symposium Remote Sensing of Urban Areas (URS 2005). Tempe, AZ, USA*, 14–16 March.

Moran, M., Jackson, R., Slater, P., and Teillet, P. (1992). Evaluation of simplified procedures for retrival of land surface reflectance factors from satellite sensor output. *Remote Sensing Environment*, **41**, 160–184.

Muchoney, D. M. and Haack, B. N. (1994). Change detection for monitoring forest defoliation. *Photogrammetric Engineering & Remote Sensing*, **60**(10), 1243–1251.

Nauck, D., Klawonn, F., and Kruse, R. (1994). *Neuronale Netze und Fuzzy-Systeme*. Vieweg, Braunschweig.

Neubert, M. and Meinel, G. (2003). Vergleich von Segmentierungsprogrammen für Fernerkundungsdaten. *Angewandte Geographische Informationsverarbeitung XV, Beiträge zum AGIT-Symposium Salzburg, Strobl/Blaschke/Griesebner (Eds.)*, 323–329.

Nielsen, A., Conradson, K., and Andersen, O. (2002). A change oriented extension of eof analysis applied to the 1996–1997 avhrr sea surface temperature data. *Physics and Chemistry of Earth*, **27**(32–34), 1379–1386.

Nielsen, A. A. (1996). Change detection in multispectral, bi-temporal spatial data using othogonal transformations. *Proceedings of the 8th Australien Remote Sensing Conference, Canberra, Australia*, March.

Nielsen, A. A. and Conradsen, K. (1997). Multivariate alteration detection (mad) in multispectral, bi-temporal data: A new approach to change detection studies. *Technical Report IMM-REP-1997.Technical University of Denmark Lyngby*, **11**.

Nielsen, A. A., Conradsen, K., and Simpson, J. J. (1998). Multivariate alteration detection (mad) and maf postprocessing in multispectral, bitemporal image data: A new approach to change detection studies. *Remote Sensing of Environment*, **64**, 1–19.

Niemeyer, I. (2001). *Satelliten und nukleare Kontrolle. Änderungsdetektion und wissensbasierte, objektorientierte Klassifikation zur Unterstützung der nuklearen Verifikation*. Schriftenreihe des Forschungszentrum Jülich, Reihe Umwelt/Environment, Band 28, Jülich.

Niemeyer, I. (2005). Use of thermal infrared image data for nuclear safeguards purposes. *Proceedings of the 27th Symposium on Safeguards and Nuclear Material Management*, 10–12 May (CD-Rom).

Niemeyer, I. and Canty, M. (2003). Pixel-based and object-oriented change detection analysis using high-resolution imagery. *Proceedings of the 25th Symposium on Safeguards and Nuclear Material Managment, Stockholm, Sweden*, 13–15 May (CD-Rom).

Niemeyer, I. and Nussbaum, S. (2005). Automation of change detection procedures for nuclear safeguards-related monitoring purposes. *Proceedings of the 31st International Symposium of Remote Sensing of Environment, Global Monitoring for Sustainability and Security, Saint Petersburg*.

Niemeyer, I., Nussbaum, S., and Lingenfelder, I. (2005a). Automated analysis of remote sensing data for extensive monitoring tasks in the context of nuclear safeguards. *Proceedings of the IEEE International Geo-science and Remote Sensing Symposium, IGARSS'05, Seoul*.

Niemeyer, I., Nussbaum, S., and Canty, M. (2005b). Automation of change detection procedures for nuclear safeguards-related monitoring purposes. *Proceedings of the IEEE International Geo-science and Remote Sensing Symposium, IGARSS'05, Seoul*.

Nussbaum, S., Niemeyer, I., and Canty, M. (2005). Feature recognition in the context of automated object-oriented analysis of remote sensing data monitoring the iranian nuclear sites. *Proceedings of the SPIE Europe Symposium on Optics/Photonics in Security & Defence, Bruges, Belgium*, September 26–28, **5988**.

Pappas, T. N. and Safranek, R. (2000). Perceptual criteria for image quality evaluation. *Handbook of image and Video Processing. In: A. C. Bovik, (Ed.)*, p. 939–959, Academic Press.

Poucet, A. (2006). Arms control and non-proliferation treaties: An ontology of concepts and characteristics. In: *Verifying Treaty Compliance – Limiting Weapons of Mass Destruction and Monitoring Kyoto Protocol Provisions*, R. Avenhaus, N. Kyriakopoulos, M. Richard and G. Stein (Eds.), Springer, New York.

Quint, F. (1998). Objekterfassung aus Karte und Bild. In H.-P. Bähr and T. Vögtle (Ed.): *Digitale Bildverarbeitung. Anwendung in Photogrammetrie, Kartographie und Fernerkundung*. Heidelberg, Germany.

Ranchin, T. and Wald, L. (2000). Fusion of high spatial and spectral resolution images: The arsis concept and its implementation. *Photogrammetric Engineering and Remote Sensing*, **66**(1), 49–61.

Ranchin, T., Aiazzi, B., Alparone, L., Baronti, S., and Wald, L. (2003). Image fusion – the arsis concept and some successful implementation schemes. *ISPRS Journal of Photogrammetry & Remote Sensing*, **58**, 4–18.

Richard, M. (2006). Beyond Iraq: The new challenges to the nuclear nonproliferation regime). In: *Verifying Treaty Compliance – Limiting Weapons of Mass Destruction and Monitoring Kyoto Protocol Provisions*, R. Avenhaus, N. Kyriakopoulos, M. Richard and G. Stein (Eds.), Springer, New York.

Richards, J. A. and Jia, X. (1999). *Remote Sensing Digital Image Analysis. An Introduction.* Springer Verlag, New york.

Richter, R. (1990). A fast atmospheric correction algorithm applied to landsat tm images. *International Journal of Remote Sensing*, **11**(1), 159–166.

Ridd, M. K. and Liu, J. (1998). A copmarison of four algorithms for change detection in urban environment. *Remote Sensing of Environment*, **63**, 95–100.

Rosenfield, G. H., Fitzpatrick-Lins, K., and Ling, H. (1982). Sampling for thematic map accuracy testing. *Photogrammetric Engineering and Remote Sensing*, **48**(1), 131–137.

Sagerer, G. and Niemann, H. (1997). *Semantic Networks for Understanding Scenes. Advanes in Computer Vision and Machine Intellegence*. Plenum Press, New York.

Sandmeier, S. (1997). Radiometrische korrektur des topographieeffektes in optischen satellitenbilddaten – vergleich eines semi-empirischen verfahrens mit einem physikalischbasierten modell. *Photogrammetrie, Fernerkundung, Geoinformation*, **1**, 23–32.

Schenkel, R. (2006). Improving verification:Trends and perspectives for research. In: *Verifying Treaty Compliance – Limiting Weapons of Mass Destruction and Monitoring Kyoto Protocol Provisions*, R. Avenhaus, N. Kyriakopoulos, M. Richard and G. Stein (Ed.), Springer, New York.

Schiewe, J. (2002). *Segmentation of High-Resolution remotely sensed Data – Concepts, Applications and Problems*. Symposium on Geospatial Theory, Processing Applications. Ottawa.

Schiewe, J. and Tufte, L. (2001). Potenzial regionen-basierter Verfahren für die intergrative Auswertung von GIS- und Fernerkundungsdaten. In Fernerkundung and GIS. T. Blascke (Eds.). Wichmann.

Schott, J., Salvaggio, C., and Volchok, W. (1988). Radiometric scene normalization using pseudo-invariant features. *Remote Sensing Environment*, **26**, 1–16.

Schott, J. R. (1997). *Remote Sensing. The Image Chain Approach.* Oxford University Press, New York.

Schowengerdt, R. A. (1997). *Remote Sensing – Models and Methods for Image Processing*. 2nd Edition, San Diego. Academic Press.

Sim, S. (2005). A proposed method for disaggregating census data using object-oriented image classification and gis. *Proceedings of the ISPRS WG VII/1 Human Settlements and Impact Analysis 3rd International Symposium Remote Sensing and Data Fusion Over Urban Areas (URBAN 2005) and 5th International Symposium Remote Sensing of Urban Areas (URS 2005). Tempe, AZ, USA,* 14–16 March.

Singh, A. (1989). Digital change detection techniques using remotly-sensed data. *International Journal of Remote Sensing*, **10**(6), 989–1003.

Slink, T. A. (1996). *Satellitenfernerkundung zur Verifkation von Rüstungskontrollen und Abrüstungsverträgen*. Europäische Hochschulschriften 1957, Reihe II, Rechtswissenschaft. Verlag Peter Lang, Frankfurt/Main.

Sohl, T. L. (1999). Change analysis in the united arab emirates: An investigation of techniques. *Photogrammetric Engineering & Remote Sensing*, **65**(4), 475–484.

Sonka, M., Hlavac, V., and Boyle, R. (1998). *Image processing, Analyses and Machine Vision, 2. Auflage*. PWS Publishing, Pacific Grove.

Sunar, F. (1998). An analysis of change in a multi-date data set: A case study in the ikitelli area, istanbul, turkey. *International Journal of Remote Sensing*, **19**(2), 225–235.

Thepaut, O., Kpalma, K., and Ronsin, J. (1998). Ers sar and spot images automatic registration in a multichannel consensual segmentation scheme. *Proceedings of the IEEE 1998 International Geoscience and Remote Sensing Symposium, Seattle, USA*, 6–10 July **II**, 1040–1042.

Tizhoosh, H. R. (1998). *Fuzzy-Bildverarbeitung. Einführung in Theorie und Praxis.* Springer, Berlin.

Tönjes, R. (1999). Wissensbasierte Interpretation und 3D-Rekonstruktion von Landschaftsszenen aus Luftbildern. *Fortschritt-Berichte VDI, VDI-Verlag. Düsseldorf*, **10**(575).

Tönjes, R., Growe, S., Bückner, J., and Liedtke, C.-E. (1999). Knowledge-based interpretation of remote sensing images using semantic nets. *Photogrammetric Engineering & Remote Sensing*, **65**(7), 811–821.

Truong, O., Keeffe, R., Baines, P., and Paquette, J.-P. (1999). Potential application of commercial satellite imagery in international safeguards. *Journal of Nuclear Materials Management (JNMM)*, **XXVII**(2), 13–18.

USGS (Januar, 1990). The spatial data transfer standard. *United States gelogical Survey, Draft.*

van Genderen, J., Lock, B., and Vass, P. (1978). Remote sensing: Statistical testing of thematic map accuracy. *Remote Sensing of Environment*, **7**, 3–14.

Vermote, E. F., Tanre', D., Deuze', J. J., Herman, M., and Mocrette, J.-J. (1997). Second simulation of the satellite signal in the solar spectrum, 6s: An overview. *Geoscience and Remote Sensing*, **35**(3), 675–686.

Wang, Z. and Bovik, A. C. (2002). A universal image quality index. *IEEE Signal Processing Letters.*

Wiemker, R. (1997). An iterative spectral-spatial bayesian labeling approach for unsupervised robust change detection on remotely sensed multispectral imagery. *Proceedings of the 7th International Conference on Computer Analysis of Images and Patterns, CAIP'97*, **10–12**, 263–270.

Wiemker, R., Speck, A. Kulbach, D., Spitzer, H., and Bienlein, J. (1997). Unsupervised robust change detection on multispectral imagery using spectral and spatial features. *Proceedings of the Third International Airborne Remote Sensing Conference and Exhibition*, 7–10 July, *Copenhagen, Denmark*, **I**, 640–647.

Wilkinson, G. G. (1999). Recent developments in remote sensing technology and the importance of computer vision analysis techniques. I., Kanellopoulos, G. G. Wilkinson, and Moons T. (Eds.): *Machine Vision and Advanced Image Processing in Remote Sensing. Berlin*, **I**, 5–11. Springer.

Witheside, T. (2005). A multi-scale object-oriented approach to the classification of multi-sensor imagery for mapping land cover in the top end. *Proceedings of NARGIS 2005 – APPLICATIONS IN TROPICAL SPATIAL SCIENCE. 4th–7th July 2005 Charles Darwin University, Darwin, NT, Australia.*

Xu, L. and Yuille, A. L. (1995). Robust principal component analysis by self-organizing rules based on statistical physics approach. *ITTT Transactions on Neural Networks*, **6**(1), 131–143.

Yang, L., Albregtsen, F., Lønnestad, T., and Grøtum, P. (1995). A supervised approach to the evaluation of image segmentation methods. *Proceedings CAIP 1995, Lecture Notes in Computer Science*, **970**, 759–765.

Yang, T.-N. and Wang, S.-D. (1999). Robust algorithms for principal component analysis. *Pattern Recognition Letters* **20**, 927–933.

Yang, X. and Lo, C. (2000). Relative radiometric normalization performance for change detection from multi-date satellite images. *Photogrammetric Engineering and Remote Sensing*, **66**, 967–980.

Zadeh, L. A. (1965). Fuzzy sets. *Information and Control*, **8**, 338–353.

Zhang, Y. (1996). A survey on evaluation methods for image segmentation. *Pattern Recognition*, **29**(8), 1335–1346.

Index

GPSR Compliance

The European Union's (EU) General Product Safety Regulation (GPSR) is a set of rules that requires consumer products to be safe and our obligations to ensure this.

If you have any concerns about our products, you can contact us on ProductSafety@springernature.com

In case Publisher is established outside the EU, the EU authorized representative is:

Springer Nature Customer Service Center GmbH
Europaplatz 3
69115 Heidelberg, Germany

Batch number: 10583998

Printed by Printforce, the Netherlands